Protein Functionality in Foods

Protein Functionality in Foods

John P. Cherry, EDITOR

Southern Regional Research Center,
USDA

Based on a symposium

sponsored by the Division

of Agricultural and Food

Chemistry at the 179th

Meeting of the American

Chemical Society, Houston,

Texas, March 24–28, 1980.

ACS SYMPOSIUM SERIES **147**

AMERICAN CHEMICAL SOCIETY

WASHINGTON, D. C.   1981

Library of Congress CIP Data

Protein functionality in foods.
  (ACS symposium series, ISSN 0097–6156; 147)

  Includes bibliographies and index.

  1. Proteins in human nutrition—Congresses. 2. Proteins—Analysis—Congresses. 3. Food—Sensory evaluation—Congresses.
  I. Cherry, John P., 1941–   . II. American Chemical Society. Division of Agricultural and Food Chemistry. III. Series.

TX553.P7P76   641.1′2   81–97
ISBN 0–8412–0605–8 AACR2 ACSMC8 147 1–332
                                      1981

ACS Symposium Series

M. Joan Comstock, *Series Editor*

FOREWORD

The ACS Symposium Series was founded in 1974 to provide a medium for publishing symposia quickly in book form. The format of the Series parallels that of the continuing ADVANCES IN CHEMISTRY SERIES except that in order to save time the papers are not typeset but are reproduced as they are submitted by the authors in camera-ready form. Papers are reviewed under the supervision of the Editors with the assistance of the Series Advisory Board and are selected to maintain the integrity of the symposia; however, verbatim reproductions of previously published papers are not accepted. Both reviews and reports of research are acceptable since symposia may embrace both types of presentation.

CONTENTS

PREFACE

Functionality, or functional property, is defined as "any property of a substance, besides the nutritional ones, that affects its utilization" (*1*). Functionality, and particularly the role of proteins in its properties, is currently a high-priority area of research. Scientists working on functionality need to understand the physicochemistry of its properties and collaborate in research on methods of testing it so that comparisons of data from different laboratories can be made and model tests of functional properties can be coordinated with actual food systems.

This book updates and presents new information on the physicochemistry of functionality, the roles and use of proteins for improving the functional properties of foods, and the application of data from model test systems to actual food ingredients. This volume should be useful to food processors, engineers, chemists, physicists, and others engaged in these or related areas of research. It also is hoped that this book will stimulate its readers to expand research on functionality.

The functional properties that govern the role of proteins in food applications are: color; flavor; texturization; solubility; viscosity; adhesion or cohesion; gelation; coagulation; aeration or foamability; water and oil absorption; and emulsification.

Visual appearance, or color, and odor, or flavor, are the first functional properties of a food that one encounters; it is these properties that lead to acceptance or rejection of a food. High-quality protein ingredients are usually light in color and have a bland flavor. Isolation and identification of the pigments that cause color in protein products are important steps in solving color or odor. Volatiles that might impart off-flavors in food ingredients can be steam distilled in situ and identified by combined gas chromatography and mass spectrometry.

Texturization is a composite of the properties that contribute to the structural elements of the ingredient, and the manner in which these constituents register with the physiological senses. Molecular size, shape, charge, and solubility, as well as the presence of carbohydrate, process temperature, and methods to enhance the interaction of these components all affect texture.

Solubility and viscosity explains the basic physicochemical nature and the functional behavior of proteins in aqueous solutions. Proteins contribute to the adhesive or cohesive properties of film matrices by the binding of their polypeptides to other components such as starch granules to yield shaped products that, for example, are necessary for trapping gases in breadmaking.

Gelation involves protein aggregation whereby polymer–solvent interactions occur as tertiary structural networks. This semielastic matrix is capable of entrapping large amounts of water and other soluble components. Most vegetable protein ingredients contain large-molecular-weight globulins that must be denatured by unfolding their polypeptides to promote molecular interactions to allow gel formation. Coagulation also involves protein aggregation but produces a less elastic, less hydrated structure than that of a gel. Heat treatment and divalent cations such as calcium are important in these reactions.

When proteins unfold and expose their hydrophobic regions, the interfacial tension between air and the aqueous phase is lowered facilitating deformation of the liquid into an impervious film that enhances aeration or foamability of the solution. This is affected by pH, ionic strength, and temperature, and protein, oil, and carbohydrate concentrations. Water and fat absorption involve similar physicochemical and environmental properties, and reportedly have been affected by the size of ingredient particles.

Emulsification is a stabilizing effect of proteins; a lowering of the interfacial tension between immiscible components that allow the formation of a protective layer around oil droplets. The inherent properties of proteins or their molecular conformation, denaturation, aggregation, pH solubility, and susceptibility to divalent cations affect their performance in model and commercial emulsion systems. Emulsion capacity profiles of proteins closely resemble protein solubility curves and thus the factors that influence solubility properties (protein composition and structure, methods and conditions of extraction, processing, and storage) or treatments used to modify protein character also influence emulsifying properties.

Unit processes that modify proteins and improve their functionality may affect the bioavailability of protein or nonprotein constituents, or both. For example, proper heat treatment increases the digestibility of proteins and destroys heat-labile, antinutritional factors, but excess heat can reduce digestibility and decrease nutritional value by destroying or altering amino acids. Another example is the reaction of succinic anhydride with the α-amino group of lysine in proteins causing changes in their molecular structure and thus altering their contribution to functionality. This reaction hinders proper absorption of lysine in the intestinal tract and lowers the nutritional value of the modified ingredient.

Proteins can be modified by proteolytic enzymes with limited reduction in their nutritional bioavailability. Enzymatic hydrolysis of peptide bonds of proteins will reduce their molecular size, affect their structures, expose different regions of their molecules to the environment, and thereby alter their contribution to functionality, e.g. by increasing and decreasing the solubility and viscosity properties, respectively, of aqueous solutions. These changes can be controlled by carefully selecting proteolytic enzymes, maintaining proper treatment conditions, and monitoring the hydrolysis reactions.

The effects of varying conditions that affect protein properties and, in turn, their contribution to functionality can be modeled by multiple regression analyses. These procedures allow the researcher to test the significance of statistics based on alternative hypotheses concerning the behavior of the functional properties. They also allow the researcher to estimate the magnitude of the effects of environmental factors in food systems on the behavior of functional properties and show the importance of experimental design in determining applicability of statistical tests.

Literature Cited

1. Pour-El, Akiva; In "Functionality and Protein Structure," *ACS Symp. Ser.* **1979,** *92.*

JOHN P. CHERRY

Southern Regional Research Center

SEA–AR, U.S. Department of Agriculture

P.O. Box 19687

New Orleans, LA 70179

October 14, 1980

1

Protein Functionality: Classification, Definition, and Methodology

AKIVA POUR-EL

PEACO, St. Paul, MN 55108

No subject in the general area of food chemistry and technology has suffered more from inconsistency, confusion, and ambiguity than the field of functionality. Even the term functionality often is applied indefinitely to some related subjects. Until now, no rigorous treatment of its coverage has emerged. Each researcher and/or author has attempted to fit the name to his/her current interests or work. The methodology has not been standardized; properties often are confused with each other and very frequently wide conclusions are drawn from data obtained by ad hoc assays of uncertain validity.

Some may imagine that this lack of rigor, standardization, and clarity, so essential to a real science, indicates a low priority for this subject and a lack of importance and of need for its methods. This is not so. It is of great importance for practical food chemistry and cognizance of this fact is growing with each year.

Enough has already been written and orated about the dangers of world hunger and the requirements for adequate nutritive foods for the multiplying billions on "spaceship Earth." Not as much emphasis has been placed on functionality, a critical element in accomplishing this task. It is, of course, axiomatic that, to accomplish its nutritive task a food must be ingested. Both human and nonhuman beings have shown consistent, distinct likes and dislikes in foods - mostly having no relation to the nutritive value, but strongly influenced by functional properties. As pointed out by Yudkin (1), nutritional properties can, at most, provide supplemental incentives to utilizing foods that are desirable because of their taste and appearance. Highly educated populations are only now trying to alter their food habits in a modest way according to nutritive dicta. But even they, being only human, would prefer these beneficial foods to possess the functional properties of their traditional favorites. It is, therefore, the duty of the food chemist to provide nutritive foods with the most desirable functional properties to ensure their wide assimilation.

0097–6156/81/0147–0001$05.00/0
© 1981 American Chemical Society

In a related area, wherever overeating has become a factor contributing to deteriorating health, nonnutritive components might be introduced into the normal diet, provided their functional properties are appealing. This is another great challenge to the food chemist. Low or noncaloric foods that imitate the present high caloric ones in functional properties could be produced. This will reduce obesity and its concomittant illnesses.

Furthermore, as will be apparent from the following sections, all properties of biological products that lead to their industrial utilization are, by definition, functional ones. The expected enhanced reliance on biological "renewable resources" due to the exhaustion of the fossil materials will necessitate increased emphasis on the clarification, evaluation, and modification of their functional properties as a prerequisite to their efficient utilization.

In this presentation, an attempt will be made to assign to functionality a distinct place within general biological properties, define specifically the terms used in the field, provide adequate classification of its components, and describe general methodological approaches. It is emphasized that all of these are proposals, set before the scientific community for appraisal, discussion, and possible alteration prior to adoption. It is hoped that this outline, combined with previous ones (2-7), will be employed in arriving at a workable scientific rigor for the field, helpful to all those involved in food chemistry and technology.

Functionality - A Biological Property

The term functionality was defined (3) for the food area a few years ago as: "Any property of a food or food ingredient except its nutritional ones that affects its utilization." On further consideration, one may substitute for food or food ingredient the term biological product. One may also divide all of the properties of biological products as: 1) Nutritional, or properties affecting the animal body after passage of food into the alimentary canal, including chemurgic, anabolic, metabolic enhancing, antimetabolic, and toxic; and 2) Functional, or properties influencing foods prior to entering the body, including enzymic, nonenzymic food functionality (sense affecting, manipulative), and industrial.

Among the nutritional properties, chemurgic and anabolic are those possessed by products ingested in large quantities for energy supply and cellular building and repair. The toxic substances might be added to the antimetabolic except for the possibility of contact toxicity of the nonmetabolic nature. The metabolic enhancers are compounds ingested in smaller quantities as aids to the chemurgic and anabolic ones.

The functional properties are divided according to a completely different pattern. Enzymic properties are functional ones because the action of the catalyst is, per se, a nonnutritional one. In certain cases, when enzymes are added to foods as in vivo digestion aids, they might be considered metabolic enhancers. Otherwise, when used externally to prepare products possessing more utile chemical and physical characteristics, the digestive nature is indubitably a functional property. It must be noted, however, that enzyme functionality usually is not studied in conjunction with the other functionalities, but is a distinct and separate branch of biochemistry involved in functional evaluation. Theoretically, a more scientific division of the functional properties could be made into molecular and non-molecular ones. (Enzymic properties would then be a division of the former.) However, traditional lines already have been set and the proposed division is closer to present research disciplines.

The nonenzymic food functionalities are separated into two factors: 1) Sense affecting (this is not sensory, which has already been appropriated for a methodological term), encompassing all those properties that influence the utilization through their perception by one or more of the five senses; and 2) Manipulative, those properties that affect the utilization of the product through easing or hindering its various process steps (at home or in the commercial plant).

Industrial properties may include some enzymic and manipulative ones but are concerned mostly with nonfood utilization.

In proposing these divisions, one must acknowledge that hybrid properties are certainly possible. Thus, an emulsifying property of a protein might be used as a sense enhancing one-- to prepare a smooth food product--or a manipulative one--to provide even subdivision for uniform administration.

Definitions

Functionality - The sum or aggregate of the functional properties of a product.

Functional Assay (or Functional Test) - Any quantitative evaluation of a functional property.

Functional Evaluation - An evaluation of the basic physicochemical properties that influence functionality. (By this definition the functional tests of enzymic properties are functional evaluations.)

Model System (or Model Test) - An evaluation of one or more functional properties that does not mimic completely the steps and ingredients of an actual food preparation.

Utility System (or Utility Test, or Food Test) - An evaluation of a functional property that mimics food preparation in all its particulars.

Functional Value - A general characterization of a product's functionality that emphasizes its utility. This is a

qualitative term and may state that a protein may have a
functional value as an emulsifier or gelling agent.

Classification of Protein Functional Properties

Many methods of classifying the functional properties of
protein products have been proposed (3). They are based on the
preferences and intentions of the researcher.

Aim of utilization. These are the main divisions of
functionality mentioned above and include sense affecting, manipulative, and industrial properties. Enzymic characteristics
are also functional properties of some protein products, but lie
outside the scope of this chapter.

Material tested. Classified as to whether the material is
a food product (including feeds) or a food ingredient. The
former's functionality affects its utilization by the consumer
while the latter is more important to the processor.

Test procedure. This divides the functional properties into
a Model System or a Utility System.

Evaluation procedure. Functionality is evaluated either by
subjective, semiobjective, or objective methods. The subjective
methods utilize exclusively the five senses as sensors and the
brain as the value assigner. Objective methods utilize completely artificial sensors and instruments as the value assigners.
The semiobjective method incorporates living elements in its
sensors but joins these to instruments for value assignment.

Physico-chemical properties. Assigns to each of these
molecular properties some functional characteristic that is
influenced by it, as follows:

Hydrophilic - properties dependent on the affinity of the
protein to water and other polar solvents, including
solubility, water uptake, and wettability.

Interphasic - properties depending on the ability of the
protein molecules to form separation and junction films
between two immiscible media, including emulsification, fat
uptake, foaming, adsorption, and coacervation.

Intermolecular - properties utilizing the ability of
proteins to form junctions of its own molecules to themselves
or to other components including viscosity, thickening, gelation, film formation, foaming, fiber formation, adhesion,
cohesion, stickiness, hardness, complex formation, spreading,
elasticity, and plasticity.

Organoleptic - properties of protein products manifested
through the sense organs including odor (nose), color (eye),
flavor (taste organs), brittleness (ear), and so forth.

Other Effects - special chemical or physical properties
that pertain to specific proteins (for example,
incompatibilities).

Protein Properties and Functionality

The factors influencing the functional properties of protein products are the innate characteristics (physico-chemical) of proteins, and processing and modification steps that alter them. Physico-chemical properties include:

Amino Acid Composition - The percentage and distribution of the amino acids has great influence on hydrophobicity, and the other factors discussed in this section. It also influences the rate of processing and modifying alterations in its nature.

Size - The larger the molecular units in the material the more insoluble it becomes. It also influences the ease of disintegration and rearrangements necessary during modification.

Conformation - The conformational characteristics are involved in functional properties through hydrophilicity and hydrophobicity, gelation, and film forming among others. Denaturation, which is a change in conformation, has always been considered the bane of protein processing prior to final form formation. At that step, controlled denaturation often is practiced.

Bonds and Forces - These properties are the mediators affecting the changes in size and conformation. Van der Waal forces, ionic bonds, hydrogen bonds, covalent bonds, and hydrophobic bonds all play a part in the original protein structure as well as in the modifications leading to altered functionality. Adequate correlations of these with functional properties are the subjects of "Functional Evaluations" (3).

Processing Factors Affecting Functionality

There are seven major process steps that usually bear some influence on the functionality of the product as follows:

Protein Source and Variety - By choosing a particular protein, one may increase a particular functionality; for example, increased sulfur amino acid concentration will enhance gelation of protein ingredients.

Extraction - Either the extraction of the protein from its source, or the extraction of other components from the protein, affects functionality.

Temperature - The denaturing effect of temperature will have a very significant effect on functionality. One must also mention the possibility of alterations through annealing (the keeping of a material at a constant temperature for a time period) on certain functionalities such as water absorption.

Drying - The methods of drying influence functionality through the temperature used. Modifications of porosity and particle size also are frequently observed during this treatment.

Ionic history - The pH history of the protein sample influences its state of naturation and its conformation. Also, the presence of particular ions will sometimes drastically influence its functionality through formation of complexes with diverse properties.

Impurities - The presence of other components such as fats, insoluble carbohydrates, sugars, and so forth, has a great effect on the functionality of the protein product. Complexing reactions may occur that severely alter the properties; (for example, grittiness resulting from insoluble components).

Storage - Storage affects the stability of a protein product to varying degrees. Even cold room storage, while apparently not detrimental to the state of naturation, may sometimes result in other chemical reactions, not all of which are beneficial. The moisture content during storage is, a well known factor affecting product characteristics.

Modification of Functional Properties

There are three main avenues for altering the functional properties of a protein product including:

Physical - These involve changes in particle size through application of temperature, removal of volatiles and moisture, and alterations in form through the application of stress.

Chemical - These depend on the alterations in the physico-chemical nature of the product through pH modifications, derivatization, enzymatic digestion, ionic manipulation, and complex formation. In these modifying steps, either the size of the molecules and/or their natures may be involved. The proteins or the impurities might be the targets of these steps.

Biological - When fermentation is used as a modifying step for altering functionality, one essentially uses sequential enzyme digestion. The main variation is that, at times, large quantities of the material are lost due to the energy needs of the microorganisms.

Foods and Functional Properties

In Table I, foods are listed showing the functional properties needed for their success.

Table I. Types of Foods and Their Related Functionalities (7)a

Type	Functionality
Beverages	Solubility, grittiness, color
Baked goods	Emulsification, complex formulation, foaming, viscoelastic properties, matrix and film formation, gelation, hardness, absorption
Dairy substitutes	Gelation, coagulation, foaming, fat holding capacity
Egg substitutes	Foaming, gelation
Meat emulsion products	Emulsification, gelation, liquid holding capacity, adhesion, cohesion, absorption
Meat extenders	Liquid holding capacity, hardness, chewiness, cohesion, adhesion
Soups and gravies	Viscosity, emulsification, water absorption
Topping	Foaming, emulsification
Whipped deserts	Foaming, gelation, emulsification

a In all of these functionalities, flavor is important.

Protein Functional Methodology - Requirements and General Design

As previously alluded to, functional methodology is not, as yet, fully standardized. The literature is bulging with methods, many times improvised on the spur of the moment to fit the needs and instruments of the researcher, with inadequate preliminary examination or statistical evaluation. This section will attempt to organize a general system of methodology for protein food products. In it, a general plan of assays will be followed by criteria for choice of assay, characteristics of Model Tests, and, finally, characteristics of Utility Tests.

Testing A Protein Product For Functional Properties.
Chemical analyses for general composition are essential to adequately do future functional testing. The percentage of protein and other components will invariably supply clues as to the expected functional properties. Physical analyses for size of

particle, appearance, and density will also aid in the search
for good functionality.

General Functional Properties. These include solubility,
emulsification, water and oil absorption, gelation or coagulation, foaming, viscosity, texture, adhesion or cohesion, and
film formation.

Special Food Tests. These may be both Model Tests and
Utility Tests. Following the application of the tests above, a
knowledgeable reasearcher will be able to assess with some certainty which of the food system Model Tests are worthwhile. It
has frequently been noticed that regardless of mediocre results
in the Model Tests, many researchers will still try to evaluate
their products in the more specialized food system Model Tests,
in which the tested product will inevitably fail. The tests
include meat emulsion system tests, extruded product tests,
baking system tests, dairy product tests, and coacervates.

Choice of Model or Utility Test. There frequently are
objections to employing any model food system test. Those who
do so insist that none have yet been designed that adequately
duplicate Utility Systems. Still, Model Tests are needed for
the following reasons:

Time - In many instances, preparation and evaluation of
Utility Tests require considerable expenditure of time.
Model Tests of shorter duration can be employed by enhancing
the severity of the test conditions.

Equipment - Some food preparing machinery is unavailable in
the smaller ingredient supplying laboratories. A Model Test
may employ smaller or simpler units immitating the processes
of larger or more complex ones.

Sample Size - Because of their mass processing nature,
commercial food systems machinery tend to utilize large
sample sizes. When novel products are tested, the experimental samples are sometimes too small to be used in such a
manner. Model Systems employing smaller batch sizes provide
the answers.

Sample Numbers - When a large number of samples has to be
tested, making a food product from each one is obviously an
impossibility. Yet the choice between them has to be made.
Model Systems provide clues as to which samples need not
undergo more extensive testing.

Number of Tests - When a sample may be useful in many
foods, a great number of Utility Tests will be required to
evaluate it fully. Model Tests allow the prescreening of
those Utility Tests that are most apt to be successful.
This is based on a general functional property.

Resolution - In certain food preparations, the tests do not
provide enough resolution to differentiate between similar
products. Yet these differences should be known when experimental products are tested as a guide for future experiments.
Also, these minor differences will appear upon standing for
long periods of time. In Model Tests, aggravated conditions

may improve the resolution and differentiate between similar products.

Availability - Some new products do not, as yet, have a utility and, therefore, a Utility Test is unavailable. A few Model Tests may guide the researcher in the planning of such Utility Tests.

Based on the above, a Model Test should be inexpensive, use simple equipment, have high resolution, and test for properties of multiple utility.

Designing a Model Test. A successful Model Test is one that leads to a minimum number of Utility Tests. To accomplish this, one has to follow the general plan leading from the first probings of the unknown product to a final estimation of its positive probable utility in a particular food system. The following steps are sequential. Skipping one of these may, at times, be possible for one who has had sufficient experience in the field. However, in general, this is not recommended; functional properties of even similar components may vary, depending on various conditions as will be specified in a later section. Much time may be lost if a logical sequence is not followed.

Preliminary Tests - These were specified previously in the text.

Choice of Physico-Chemical Properties - This is based on the general functionality tests enumerated previously. Some of these may be omitted by wise deductions from the general physico-chemical properties of similar products. Final selection of a useful physico-chemical property is dependent not only on the actual properties of the product tested but may also be influenced by economic and market factors.

Choice of Utility - In this step, market knowledge is predominant. No one need test by a Model Test a product that is intended to supplant a presently abundant, inexpensive material. The advantages of the tested product have to be assessed clearly before it is considered as worthy of taking on such competition. We have noted frequently that experimental products are proposed as functionally useful, merely because their properties barely approach those of commonly accepted products. No attempt is made to justify their use on any economic basis. This is to be deplored. An abundant supply of the material may sometimes justify this approach but only if consideration is given to the cost of bringing it to market in an actually useful form. When harvesting, processing, and marketing costs are added, many times a "cheap" product ceases to be cheap.

Novel utilities for presently available products are a more productive area. (Who would not be interested if the fiber from grains could be used functionally in the snack food industry for reasons other than as a filler.) But these types of utility demand extreme physico-chemical modifications involving prolonged research and much expenditure.

Model Mixture Formulations - Here the knowledge of the chemist and food technologist are strongly brought to bear on the testing. The main purpose is the imitation of food systems or food formulations presently accepted. Knowledge of which components in such formulations are essential to the functionality is of utmost importance. Elimination of unnecessary and costly ingredients will not only make the test less expensive but may sometimes eliminate the conditions leading to ambiguity and reduced resolution.
Stress Conditions - These are added to improve the resolution of the test. These stress steps are dependent on processing and storing conditions. In most tests, a normal process step is increased in severity of duration. Similarly, accelerated storage steps are employed to put a stress on the stability of the mixture.
Products Tested - The Model Test is now employed in testing not only the product under consideration, but also similar market products of known success as well as products known to have failed in such a utility, if any can be found. The duplication of the process steps, employed in producing the new product on the successful market products, may sometimes be a preliminary step to this test. If they fail any of these steps, alternate processing may be needed prior to the assay. In choosing competitive product testing, care should be taken to select those resulting in numerical values similar to the materials tested both above and below it. Widely separated results indicate a sloppy approach and unnecessary efforts.
Sensory Test - The choicest of the tested materials are now subjected to an actual Utility Test in a food system. These will include both the successes and the failures of the Model Test, with a range of values of significant variance. The foods prepared are then tested by a purely sensory approach.
Correlation - Finally the Model Test results are correlated with the sensory tests. Adequate statistical tools are necessary of which linear correlations are not always the desired ones when comparing sensory values with objective test results. Only when this final correlation is achieved has an adequate Model Test been devised.
Utility tests. By definition, these are actual food formulations copying accepted food preparations. As mentioned above, when Model Tests are designed, the range of products undergoing the test should be wide enough to include some failures. In some cases, a Utility Test is employed that substitutes objective evaluation systems for sensory ones. This reduces the time of the test and its cost. However, only those objective tests previously found to be well correlated with sensory tests should be employed. Eventually a sensory test has to be performed. No food product should be marketed without a final utility test employing subjective evaluations.

Protein Functional Methodology - Outline of General M(

In a previous section some general functional properties that are commonly looked for in protein products were classified according to physico-chemical properties. In this section, the general methodology for the objective evaluation (excluding sensory subjective methods), of these properties will be outlined.

General methods of testing. Some general principles involved in sample preparation and the test array are described in the methods outlined below:

Subdivision - In most cases, the product is first subjected to subdivision to a standard size. (In texture evaluation, in some cases, this is not included.) It is generally supposed that this process will improve the contact of the product with other assay components and will reduce the hazards of unequal distribution and localized effects. The particle size has a strong influence on the value obtained in the test.

Mixing - This step sometimes includes a pre-exposure step where the tested product is allowed to equilibrate with the testing medium. The severity of the mixing (presence of shear) and its duration will affect the test results, not only through the heat generated but also through specific molecular affects such as denaturation.

Temperature - Protein molecules are very susceptible to heat effects. Therefore, most tests have to be defined clearly as to these conditions. In many traditional tests still used, adequate control of the temperature is not attempted. This leads to the frequently observed discrepancies in the literature.

Ionic Medium - Since many of the properties of protein products depend on their ionic configuration and amphionic nature, changes in pH will strongly influence the results. Standardized media are to be recommended, or preferably, correlate the results with pH changes in the medium. Similarly, the presence of other ions needs to be either clearly stated or excluded.

Stress Conditions - In many cases stress conditions are employed either prior to actual testing or as part of the measurement procedure itself. Needless to say, the standardization of these stress conditions or their absence will greatly modify the test results.

Solubility measurement. The general scheme of testing for solubility involves subdivision, exposure, agitation, separation, and measurement ($\underline{4}$, $\underline{8}$-$\underline{21}$).

The severity of agitation and separation determines whether the separated medium is considered to have proteins in solution (protein solubility values) or in dispersion (protein dispersion values). The terms Nitrogen Solubility Index (NSI) and Protein Dispersibility Index (PDI) are to be used only when the official

oyed. Any deviation from these proce-
_, or equipment modification) should be
/ other terms such as Protein Solubility
/eferably, Protein Solubility Value (PSV)
 bility Value (PDV). The process of separa-
ds is usually through application of force
_y) or filtration.
_ - This name can be used generically to charac-
tc   properties involving interaction between the
prote   /roduct and water as a result of which some of the
water remains with the product (15, 22-29). The
hydrophillic attraction is manifested in the quantity taken
up as in sorption (or true hydration) and swelling (increase
in product volume). It can also manifest itself in the
quantity staying with the product after exposure to excess
water as in water binding, water retention, or water holding capacity. Another aspect of this attraction is that of
wettability, or the rate of water uptake. The uniform steps
for these tests involve exposure, separation, and measurement. Sometimes agitation is added following exposure and
subdivision prior to it. The severity of separation is of
special significance in these methods and has a great
influence on the results.

Interphase properties. These properties can best be
explained by their relationships within the general family of
colloids (Table II).

Emulsification. Emulsions are dispersed immiscible droplets
within another liquid stabilized by the interphasic compounds
(30-49). In dealing with proteins, the disperse liquid is a fat
or oil and the stabilizing interphase is a protein product; in
butter, these phases are reversed.

Emulsions can generally be divided into three main classes
depending on consistency: thin, thick, or solid.

Emulsion Capacity is the property of the protein product
solution or suspension to emulsify oil. The measurement is of
the maximum amount of oil that the mixture will emulsify without
losing its emulsion characteristics. The steps involved in this
test are: 1) Hydration - formation of the aqueous mixture. 2)
Oil addition - with agitation the cause of emulsification. 3)
Stress - a result of the heat generated during emulsification.
4) Measurement the breakpoint (visual, sonic, electronic, etc.)
and the amount of oil emulsified (capacity) to the breakpoint
(volumetric, gravimetric, viscometric, etc.).

Protein Functional Methodology - Outline of General Methods

In a previous section some general functional properties
that are commonly looked for in protein products were classified
according to physico-chemical properties. In this section, the
general methodology for the objective evaluation (excluding sensory subjective methods), of these properties will be outlined.

General methods of testing. Some general principles
involved in sample preparation and the test array are described
in the methods outlined below:

Subdivision - In most cases, the product is first subjected
to subdivision to a standard size. (In texture evaluation,
in some cases, this is not included.) It is generally supposed that this process will improve the contact of the
product with other assay components and will reduce the
hazards of unequal distribution and localized effects. The
particle size has a strong influence on the value obtained
in the test.

Mixing - This step sometimes includes a pre-exposure step
where the tested product is allowed to equilibrate with the
testing medium. The severity of the mixing (presence of
shear) and its duration will affect the test results, not
only through the heat generated but also through specific
molecular affects such as denaturation.

Temperature - Protein molecules are very susceptible to
heat effects. Therefore, most tests have to be defined
clearly as to these conditions. In many traditional tests
still used, adequate control of the temperature is not
attempted. This leads to the frequently observed
discrepancies in the literature.

Ionic Medium - Since many of the properties of protein
products depend on their ionic configuration and amphionic
nature, changes in pH will strongly influence the results.
Standardized media are to be recommended, or preferably,
correlate the results with pH changes in the medium.
Similarly, the presence of other ions needs to be either
clearly stated or excluded.

Stress Conditions - In many cases stress conditions are
employed either prior to actual testing or as part of the
measurement procedure itself. Needless to say, the standardization of these stress conditions or their absence will
greatly modify the test results.

Solubility measurement. The general scheme of testing for
solubility involves subdivision, exposure, agitation, separation,
and measurement (4, 8-21).

The severity of agitation and separation determines whether
the separated medium is considered to have proteins in solution
(protein solubility values) or in dispersion (protein dispersion
values). The terms Nitrogen Solubility Index (NSI) and Protein
Dispersibility Index (PDI) are to be used only when the official

methods so named are employed. Any deviation from these procedures (such as pH, ionic, or equipment modification) should be
signified by the use of other terms such as Protein Solubility
Index (PSI) or even preferably, Protein Solubility Value (PSV)
and Protein Dispersibility Value (PDV). The process of separation in these methods is usually through application of force
(be it only gravity) or filtration.

Water Uptake - This name can be used generically to characterize all properties involving interaction between the
protein product and water as a result of which some of the
water remains with the product (15, 22-29). The
hydrophillic attraction is manifested in the quantity taken
up as in sorption (or true hydration) and swelling (increase
in product volume). It can also manifest itself in the
quantity staying with the product after exposure to excess
water as in water binding, water retention, or water holding capacity. Another aspect of this attraction is that of
wettability, or the rate of water uptake. The uniform steps
for these tests involve exposure, separation, and measurement. Sometimes agitation is added following exposure and
subdivision prior to it. The severity of separation is of
special significance in these methods and has a great
influence on the results.

Interphase properties. These properties can best be
explained by their relationships within the general family of
colloids (Table II).

Emulsification. Emulsions are dispersed immiscible droplets
within another liquid stabilized by the interphasic compounds
(30-49). In dealing with proteins, the disperse liquid is a fat
or oil and the stabilizing interphase is a protein product; in
butter, these phases are reversed.

Emulsions can generally be divided into three main classes
depending on consistency: thin, thick, or solid.

Emulsion Capacity is the property of the protein product
solution or suspension to emulsify oil. The measurement is of
the maximum amount of oil that the mixture will emulsify without
losing its emulsion characteristics. The steps involved in this
test are: 1) Hydration - formation of the aqueous mixture. 2)
Oil addition - with agitation the cause of emulsification. 3)
Stress - a result of the heat generated during emulsification.
4) Measurement the breakpoint (visual, sonic, electronic, etc.)
and the amount of oil emulsified (capacity) to the breakpoint
(volumetric, gravimetric, viscometric, etc.).

Table II. Colloidal Systems

Disperse Phase	Continuous Phase	Common Names	Examples
Solid	Liquid	Sol	Proteins in water solution
Liquid	Liquid	Emulsion	Milk, Mayonnaise
Gas	Liquid	Foams	Whipped toppings
Liquid	Solid	Gels, Solid emulsions	Jellies, Meat products
Gas	Solid	Open structures	Baked goods, Textured products

Emulsion Stability is the property of the emulsifier (the interphase molecules) to stabilize an emulsion following its formation and, sometimes, following certain stress conditions. In this test, only three steps are observed: emusification, stress, and measurement. The stress may be only an applied force (gravity or centrifugal forces) or heat or a combination of both.

The methodological factors having a special influence on these tests of emulsifying properties are subdivision methods for the colloidal system, ratio of components (water: protein: oil), and the nature of the fat used. The second one is the most neglected of these influences and is usually chosen by the researcher at will.

Fat attraction. These are methods originally designed for only a two phase system, that is, the material and the oil. They now include the model systems immitating solid emulsions that contain large quantities of water and aqueous particulates (50-55).

Fat absorption and binding properties are determined by mixing food ingredients with fats and oils, and after a separation step (centrifugation) the amount of oil absorbed is measured. The test is similar to water uptake, that is, the method includes steps of subdivision, agitation, separation, and measurement.

Fat retention capacity should be used only for the cases where aqueous preparations of the materials are treated with fats or oil (as in some food preparations), stress applied (usually by heating), and the separated oil measured. A common test of this nature is one used on meat emulsion systems following their cooking process. The special methodological influences strongly affecting these tests are component ratios, the nature of the fat, and the measurement techniques.

Aeration. This interphasic property of protein products, also called foaming or whipping, depends on the ability of the protein to form protective films around gas bubbles. Coalescence, and subsequent break-up of the membrane-like system around the bubbles by the proteins is thereby prevented (55-60).

Foaming capacity is a measure of the ability of the protein to form a gas-filled cellular system under specified conditions; the experimental conditions have not as yet been clearly standardized. In practice, aeration involves solution, air entrapment and measurement. The latter may be expressed as percentages of volume of foam, or volume increase. Also used are foam viscosity, density, and/or the energy input. Foam stability is a measure of the ability of the product to keep a foam, once formed, from collapsing. Again, neither the system of foam formation nor the conditions of the stability test have been standardized. The latter involves some stress conditions (sometimes only the force of gravity on the foam) and measurement techniques such as final volume, rate of volume change, or weight or volume of the drip. The special methodological influences are aeration and measurement. The shear involved in foam formation has to be well balanced to provide enough force to cause partial film denaturation but not to exceed the balance of the system.

Intermolecular interaction measurements. The functional properties are dependent on the protein molecules interacting with others of their own kind or with dissimilar ones (24, 51, 61-65).

Viscosity - In discussing this property we include not only viscosity but fluidity and thickening in heavy suspensions. Tools used in measuring viscosity vary from sophisticated equipment such as recording viscometers and plasticorders to simpler capillary viscometers, cup viscometers, falling ball units, or even simple test tubes affording but semi-qualitative values. The special methodological factors include the protein concentration, especially in thermo-tropic mixtures, temperature, which in this case severely alters the values, and the systems of measurement, which cannot be substituted for each other without severe discrepancies in obtained values.

Matrix and Film Formation - In this category, we include gelation, and fiber, film, and dough formation (34, 61, 64, 66-78). In all of these properties, the process involves the ability of the protein to form intermolecular bonds of a stable nature to yield a distinct semisolid form that is sometimes converted to a more solid form by further processing. General methodology includes the hydration and dispersion of the product in an aqueous medium, setting, by temperature or other means, applying stress to the form, and measurement. The latter usually employs tools of measuring resistance of the matrix to deformations, for example, viscometers, gelometers, penetrometers, stress or strain recorders. Some novel microscopic methods have also been developed. The special methodological factors are the hydration capability, the setting means, especially temperature, the stress applied during measurement, and the impurities present.

Solid Texture Properties - These properties often are a result of further treatment of the matrix and film as discussed in the previous section. They have to be treated separately because of the distinct methods of their assays (79-91).

General methodology includes preparation of the solid texture. The measurement techniques may be the application of forces such as pure stress (also strain), shear, or their combinations and the measurement of the resistance to this procedure (Table III). Also used are special methods such

Table III. Measurable Properties of Texture

Method	Property Measured
Stress (Strain)	Hardness, plasticity, springiness, elasticity (juiciness?)
Shear	Cohesion, adhesion, juiciness
Visual	Combined properties of unknown porosity nature
Others	Resistance to agents, hardness, porosity

as visual and microscopic inspection, density measurements of the product, and the ease of its aqueous disintegration (wet screening).

The properties measured in these methods are very diverse. Only recently have some of them been defined clearly in relation to the methodology involved and correlation obtained between the objective analyses and subjective tests (82).

CONCLUSIONS

The specific realm of food functional properties can be identified clearly as a part of the general area of biological properties. Some definition of terms used in it was attempted. There is a clear relationship between these properties and the physico-chemical attributes of compounds in foods. Also, one can generally find a correlation of functional properties and the classes of food preparations. The methods for obtaining clear, unambiguous values for such properties have been outlined. There are many pitfalls awaiting those engaged in this field and further standardization and codification are needed.

Literature Cited

1. Yudkin, I. Proc. 1st Int'l Congr. Environment.
     Biol. Med. UNESCO, Paris, France, 1974.
2. Pour-El, A. In "World Soybean Research," Hill, L. D.,
     ed., Interstate, Danville, Il., 1976, p. 918.
3. Pour-El, A. In "Functionality and Protein Structure,"
     Pour-El, A., ed., Amer. Chem. Soc. Symposium
     Series 92, Washington, D.C., 1979, p. ix.
4. Kinsella, J. E. J. Amer. Oil Chem. Soc., 1979, 54, 242.
5. Wolff, W. J. J. Agr. Food Chem., 1970, 18, 969.
6. Wolff, W. J.; Cowan, J. C. CRC Crit. Rev. Food Technol.,
     1971, 2, 81.
7. Kinsella, J. E. CRC Crit. Rev. Food Sci. Nutrit.
     1976, 7, 219.
8. American Association of Cereal Chemists. "Approved
     Methods," St. Paul, Minnesota, 1979.
9. American Oil Chemists Society. "Official and
     Tentative Methods," Link, W. E., ed., Champaign,
     Illinois, 1979.
10. Cherry, J. P.; McWatters, K. H. J. Food Sci., 1975,
     40, 1257.
11. Cherry, J. P.; McWatters, K. H.; Holmes, M. R.
     J. Food Sci., 1975, 40, 1199.
12. Fukushima, D. Bull. Agr. Chem. Soc. Jap., 1959, 23, 15.
13. Gilberg, L. J. Food Sci., 1978, 43, 1219.
14. Hang, Y. D.; Steinkraus, K.; Hackler, L. R.
     J. Food Sci., 1970, 35, 318.
15. Hutton, C. W.; Campbell, A. M. J. Food Sci.,
     1977, 42, 457.
16. McWatters, K. H.; Cherry, J. P. J. Food Sci.,
     1977, 42, 1444.
17. Shen, J. L. Cereal Chem., 1976, 53, 902.
18. Shibasaki, K.; Kimura, Y.; Okuba, K.; Takahashi, K.;
     Sasaki, M. J. Food Sci. Technol. (Japan), 1972, 19, 96.
19. Smith A. K.; Rackis, J. J.; Isnardi, P.; Carter, J. L.;
     Krober, O. A. Cereal Chem., 1966, 43, 261.
20. Wu, Y. V. J. Agric. Food Chem., 1978, 26, 305.
21. Yasumatsu, K.; Toda, J.; Kajikawa, M.; Okamoto, N.;
     Nori, H.; Kriwayama, M.; Ishii, K. Agric. Biol.
     Chem., 1972, 36, 523.
22. Berlin, E. P.; Kliman, P. G.; Pallansch, M. J.
     J. Colloid Interface Sci., 1970, 34, 488.

23. Carlin, F.; Ziprin, Y.; Zabik, M. E.; Kragt, L.;
    Poisiri, A.; Bowers, J.; Rawley, B.; Van Duyne, F.;
    Perry, A. K. J. Food Sci., 1978, 43, 830.
24. Fleming, S. E.; Sosulski, F. W.; Kilara, A.; Humbert,
    E. S. J. Food Sci. 1974, 39, 188.
25. Hagenmaier, R. J. Food Sci., 1972, 37, 965.
26. Hermansson, A.-M. Lebensmitte. Wissensch. Technol.,
    1972, 5, 24.
27. Johnson, D. W. Food Prod. Devel. 1969, 3 (8), 78.
28. Leung, H. K.; Steinberg, M. P.; Wei, L. S.;
    Nelson, A. I. J. Food Sci. 1976, 41, 297.
29. Miller, H. Cereal Chem., 1968, 34, 109.
30. Acton, J. C.; Saffle, R. L. J. Food Sci., 1970, 35, 852.
31. Borton, R. J.; Webb, N. B.; Bratzler, L. J. Food
    Technol., 1968, 22, 506.
32. Crenwelge, D. D.; Dill, C. W.; Tybor, P. T.; Landmann,
    W. A. J. Food Sci., 1974, 39, 175.
33. Hegert, G. R.; Bratzler, L. J.; Pearson, A. M.
    J. Food Sci., 1963, 28 663.
34. Hermansson, A.-M. J. Amer. Oil Chem. Soc., 1979, 56, 272.
35. Inklaar, D. A.; Fortuin, J. Food Technol., 1969, 23, 103.
36. Kamat, V. B.; Graham, G. E.; Davis, M. A. F. Cereal Chem.
    1978, 55, 295.
37. Pearce, K. N.; Kinsella, J. E. J. Agric. Food Chem.,
    1978, 26 716.
38. Pearson, A. M.; Spooner, M. E.; Hegarty, G. R.; Bratzler,
    L. J. Food Technol., 1965, 19, 1841.
39. Puski, G. Cereal Chem., 1976, 53, 650.
40. Saffle, R. L. Adv. Food Res., 1968, 16, 105.
41. Schut, J. Die Fleischwirtsch., 1968, 48, 1029.
42. Smith, G. C; John, H.; Carpenter, Z. L.; Mattil, K. F.;
    Cater, C. M. J. Food Sci., 1973, 38, 849.
43. Swift, C. E.; Lockett, C.; Fryr, A. J. Food Technol.,
    1961, 15, 468.
44. Tornberg, E.; Lundh, G. J. Food Sci., 1978, 43, 1553.
45. Tomberg, E. J. Food Sci., 1978, 43, 1559.
46. Titus, T. C.; Wianco, N. W.; Barbour, H. F.;
    Mickle, J. B. Food Technol., 1968, 22, 1449.
47. Tsai, R.; Cassens, R. G.; Briskey, E. J. J. Food Sci.,
    1972, 37, 286.
48. Webb, N. B.; Ivey, F. J.; Jones, V. A.; Monroe, J. R.
    J. Food Sci., 1970, 35, 501.
49. Yasumatsu, K.; Sawada, K.; Moritaka, S.; Misati, M.;
    Toda, J.; Wadci, T.; Ishii, K. Agric. Biol. Chem.,
    1972, 36, 719.
50. Anderson, R. H.; Link, K. D. Food Technol., 1975,
    29, 44.
51. Beuchat, L. R.; Cherry, J. P.; Quinn, M. R. J. Agric.
    Food Chem., 1975, 23, 616.
52. Childs, S. A.; Park, K. K. J. Food Sci., 1977, 41, 713.

53. Judge, M. D.; Haugh, C. G.; Zachariah, G. L.; Parmelee,
    C. E.; Pyle, R. L. J. Food Sci. 1974, 39, 137.
54. Satterlee, L. D.; Zachariah, N. Y.; Levin, E. J. Food Sci.,
    1973, 38, 268.
55. Berry, R. E.; Bisset, O. W.; Lastinger, J. C.
    Food Technol., 1965, 19, 1168.
56. Eldridge, A. C.; Wolff, W. J.; Nash, A. M.; Smith, A. K.
    J. Agric. Food Chem. 1963, 11, 323.
57. Lawhon, J. T.; Cater, C. M.; Mattil, K. F. J. Food Sci.,
    1972, 37, 317.
58. McWatters, K. H.; Cherry, J. P. J. Food Sci. 1976, 42,
    1205.
59. Puski, G. Cereal Chem., 1975, 52, 655.
60. Sekul, A. A.; Vinnett, C. H.; Ory, R. L. J. Agric.
    Food Chem., 1978, 26, 855.
61. Finney, E. E. Food Technol., 1972, 26, 68.
62. Fleming, S. E.; Sosulski, F. W.; Hamon, N. W. J. Food Sci.,
    1975, 40, 805.
63. Griffith, D. L.; Rao, J. N. M. J. Food Sci. 1978, 43, 531.
64. Huang, F.; Rha, C. K. J. Food Sci. 1971, 36, 1131.
65. Tung, M. A. J. Text. Stud., 1978, 9, 3.
66. Aoki, H. Nippon Nogei Kagaku Kaishi, 1965, 39, 262.
67. Aoki, H. Nippon Nogei Kagaku Kaishi, 1965, 39, 277.
68. Aoki, H.; Sakurai, M. Nippon Nogei Kagaku Kaisi, 1969,
    43, 448.
69. Cumming, D. B.; Tung, M. A. J. Can. Inst. Food Sci.
    Technol., 1977, 10, 109.
70. Evans, L. G.; Volpe, T.; Zabik, M. E. J. Food Sci.
    1963, 11, 323.
71. Fleming, S. E.; Sosulski, F. W. Cereal Chem. 1978, 55, 373.
72. Kelly, J. J.; Pressey, R. Cereal Chem. 1966, 43, 204.
73. Matthews, R. H.; Sharpe, E. J.; Clark, W. M. Cereal Chem.
    1970, 47, 181.
74. Mattil, K. F. J. Amer. Oil Chem. Soc. 1971, 48, 477.
75. Pour-El, A.; Swenson, T. S. Cereal Chem. 1976, 53, 438.
76. Suzuki, T.; Fujiwara, T. Canner's J. (Japan), 1973,
    52, 772.
77. Voisey, P. W.; Rasper, V. F.; Stanley, D. W.
    "Rheology and Texture in Food Quality." AVI,
    Westport, Conn., 1976.
78. Yasumatsu, K.; Toda, J.; Wada, T.; Misaki, M.; Ishii, K.
    Agric. Biol. Chem. 1972, 36, 537.
79. Aquilera, J. M.; Kosikowski, F. A. J. Food Sci. 41,
    647, 1976.
80. Anderson, Y. Proc. IV Int. Cong. Food Sci. Technol.,
    1974, p. 332.
81. Breene, W. M. J. Text. Stud., 1975, 6, 53.
82. Breene, W. M. J. Text. Stud., 1978, 9, 77.
83. Bourne, M. C. Texture Properties and Evaluations of
    Fabricated Foods. In "Fabricated Foods. G. Englett,
    ed., AVI, Westport, Conn., 1974, p. 127.

84. Cegla, C. F. J. Food Sci., 1978, 43, 775.
85. Cumming, D. B.; Stanley, D. W.; Deman, J. M. J. Can. Inst.
     Food Sci. Technol., 1972, 5, 124.
86. McWatters, K. R. J. Food Sci., 1977, 42, 1492.
87. Peleg, M. J. Food Sci., 1978, 43, 1093.
88. Saio, K. J. Text. Stud., 1978, 9, 159.
89. Soo, H. M.; Sander, E. H. J. Food Sci., 1977, 42, 1522.
90. Stanley, D. W.; Deman, J. M. J. Text. Stud., 1978, 9, 59.
91. Taranto, M. V.; Cegla, C. F.; Bell, K. R.; Rhee, K. C.
     J. Food Sci., 1978, 43, 767.

RECEIVED October 21, 1980.

Color

F. A. BLOUIN, Z. M. ZARINS, and J. P. CHERRY

Southern Regional Research Center, Science and Education Administration,
Agricultural Research, U.S. Department of Agriculture,
P.O. Box 19687, New Orleans, LA 70179

The first encounter with any food product is usually visual.
Immediately, a judgment is made on the appearance of the
product. The individual making the evaluation is probably not
consciously aware of the complex factors that influence this
judgment. Color is usually the major criterion used to evaluate
quality, and past experience plays a role in this evaluation.
For example, the color of fresh fruit indicates ripeness and the
time the fruit is most likely to taste best. Off-colors in
cheese or meats are associated with poor flavor quality. Generally, we expect each food product to have a certain color.
Deviations from this expected color can result in rejection of a
product even when this color does not adversely influence the
flavor or nutritional value of the food. If plant-protein
products are to attain widespread utilization in food applications, the color they impart to the product must be considered
an important factor in consumer acceptability.

Cottonseed flour is a plant protein product which can be
used in foods to improve nutritional and functional properties.
However, in some applications, cottonseed flour causes a serious
color problem. Biscuits prepared with 20% cottonseed flour are
yellow brown (Figure 1). Liquid-cyclone-processed (LCP) cottonseed flour is made from glanded seeds by liquid centrifugation
in hexane to remove the pigment glands containing the toxic
gossypol pigment (1). LCP flour produced a much darker color in
biscuits than hexane-defatted flour made from glandless
cottonseeds.

Undesirable colors in raw and processed foods have often
been minimized by 1) altering their processing to include solvent extraction steps to remove pigments, 2) adding antioxidants
or bleaching reagents to reduce color changes, or 3) including
additives to mask undesirable color. These approaches to the
color problem in cottonseed flours have been generally unsuccess-
ful (2, 3). The objective of the present research is to isolate
and identify the pigments responsible for the color problem that
occurs when cottonseed flours are used as food ingredients.
This basic approach has led to a better understanding of the

problem. The methods and techniques developed in this study
should also be applicable to defining the color problem of other
plant-protein materials, including sunflower and alfalfa leaf
proteins.

Experimental Procedures

Cottonseed flours and methods used to fractionate, isolate,
and identify pigments in these flours are fully described by
Blouin and Cherry (4) and Blouin, et al. (5).

Sunflower and soy flours were obtained from the Food Protein
R & D Center, Texas A&M University, College Station, Tex.;
alfalfa leaf protein from the Western Regional Research Center,
U.S. Department of Agriculture, Berkeley, Calif.; and peanut
flour from Gold Kist, Inc., Lithonia, Ga.

Biscuits were prepared from plant-protein flours based on
20% replacement for wheat flour. In biscuits prepared with fractions isolated from cottonseed flours, the quantities used were
calculated using the percentages these fractions represented of
the original flour, e.g., the salt solution soluble fraction of
LCP flour was 43% of the original flour and 8.6%, (20 x 0.43)
replacement for wheat was used to prepare the biscuit.

Solvent systems used for thin layer chromatography were 1)
n-butanol:acetic acid:water (4:1:5 upper phase), 2) acetic acid:
water (15:85), 3) ethyl acetate:pyridine:water (12:5:4), and 4)
chloroform:acetic acid:water (50:45:5). Silica gel plates were
used for chromatography of flavonoid aglycones and cellulose
plates for all other components. Aluminum chloride was used for
detection (under long UV light) of flavonoids, aniline phthalate
for sugars, ninhydrin for amino acids and iodine for other components. Cellulose thick layer plates were developed with
solvents 1 or 2.

Column chromatography of flour extracts was carried out on
Sephadex G-15 with water as the eluent, Sephadex LH-20 with 50%
methanol as the eluent for flavonoids, and DMF as the eluent for
gossypol components.

Flavonoid identification methods were those of Mabry et al.
(6).

Color Measurements

In any study of food color, one of the basic problems is
measuring color and reporting the results in a realistic and
understandable manner. Color is not just a physical property of
the object. Color is a sensory property dependent on both
physical and psychological factors related to the object, the
conditions of observation, and the individual making the observation. Because of the complexity and great economic importance
of the color phenomena, a vast "color science" literature that
deals with measuring and specifying color has evolved during the

past 20-30 years (7, 8). If the incident light and the observer
are removed as variables by using appropriate internationally
accepted standards, then the problem of color measurement and
specification is reduced to measuring the light-reflecting or
light-transmitting properties of the colored object. To define
the color of an object quantitatively, three fundamental quantities must be specified. They are: 1) hue or spectral color,
which identifies the object as red, green, yellow, blue, or an
intermediate color between these, 2) saturation or purity, which
is the strength or intensity of the hue, and 3) lightness or
luminance, which is the amount of light reflected or transmitted
from the object. Numerous visual and instrumental methods and
color scales have been devised to measure and express these
three quantities.

In the work with cottonseed flours, we used the Hunterlab
color meter D25D2A and expressed these measurements as Hunter L,
a, b color values. These are coordinates of the three-dimensional opponent-color space shown in Figure 2. The L value
measures lightness, or the amount of light reflected or transmitted by the object. The a and b values are the chromaticity
coordinates from which information about hue and saturation can
be obtained. The a value measures redness when plus and green-
ness when minus. The b value measures yellowness when plus and
blueness when minus.

Comparison of color measurements on wheat flour and six
different plant-protein products with the visual appearance of
these flours shows that the L, a, b scale does yield meaningful
values (Figure 3). The L value for the alfalfa leaf protein (L
= 75.5) shows that this product is much darker than the wheat
flour (L = 89.5) and somewhat darker than the other plant-
protein materials (L = 79.5 to 88.2). The a values do not show
any significant variations (a = +1.0 to -1.7), indicating the
absence of significant red or green coloration in the flours.
All of the plant-protein flours gave positive b values (b = 7.6
to 13.8), indicating some yellow coloration in these products.
This comparison of wheat and plant-protein products as dry
powders or flours also illustrates another important point: the
color of the dry flours does not accurately reflect the magnitude of the color problem observed when these flours are used in
food products.

Color measurements made on wheat and LCP cottonseed flour as
dry powders, aqueous pastes, and alkaline pastes are shown in
Figure 4. The L and b values for aqueous and alkaline (pH 10)
pastes of LCP flour, when compared to the values for wheat flour
pastes, more adequately reflect the magnitude of the color problem of the cottonseed flour. The aqueous paste is a dark
yellow brown, and the darkness and yellowness are even more
pronounced in the alkaline paste.

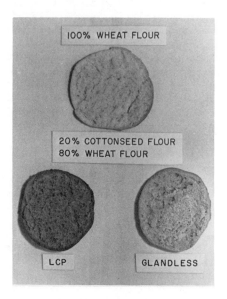

Figure 1. Biscuits containing 100% wheat and 20% LCP and glandless cottonseed flours

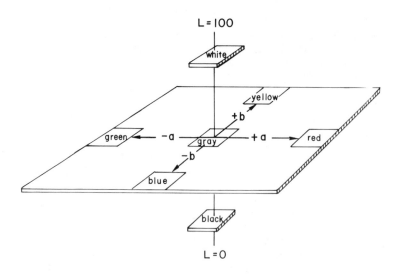

Figure 2. L, a, b color solid

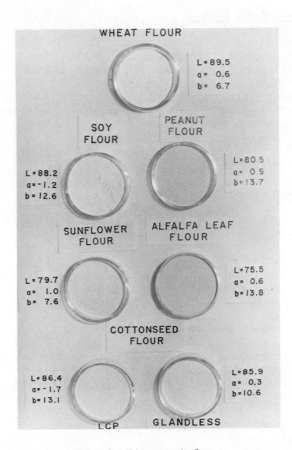

Figure 3. Plant protein flours

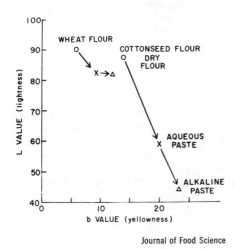

Figure 4. Hunter L and b values of dry flours (○), aqueous pastes (×), and alkaline pastes (△) of LCP cottonseed and wheat flours (4)

Biscuits as a Model System

The best method for evaluation of the color problem is to prepare a food product containing the protein flour. Biscuits were routinely used as the model food system. Figure 5 illustrates the color of biscuits prepared with 100% wheat flour and with 20% plant-protein products. The color of the biscuits prepared with soybean and peanut flours shows that these ingredients do not cause a serious color problem. However, sunflower, alfalfa leaf, and cottonseed flours do produce a discoloration in this model food system. The L and b values generally reflect this visual evaluation.

Plant Phenols

Most brightly colored plant parts contain anthocyanin, carotenoid, or chlorophyll pigments. These pigments, however, do not seem to play a significant role in the color problem associated with products containing oilseed-protein. Plant phenols, on the other hand, seem to be the most important contributors to the color problem of these protein materials. Phenols are substrates for enzymatic and nonenzymatic browning reactions. Oxidative reactions of phenol quite often lead to attachment of these compounds to protein and polysaccharide components of the plant-protein products. The role of chlorogenic acid as the source of the color problem in sunflower-seed-protein products has been reported (9, 10, 11). In cottonseed flours (4, 5), two types of phenolic compounds involved in the discoloration problem are: gossypol and flavonols.

Cottonseed Flour Fractionation

LCP cottonseed flour was separated into four fractions by methods developed for protein isolation (4). The flour was first extracted with water to yield a water-soluble and a water-insoluble fraction. The water-soluble fraction was dialyzed in cellulose tubing with a molecular weight cutoff of 3,500 to give low-molecular-weight (L) and high-molecular-weight (H) fractions. The water-insoluble part of the flour was extracted with 10% sodium chloride solution to produce a soluble (S) and an insoluble (I) fraction. Biscuits containing these fractions in the percentages that were present in the original flour were prepared (Figure 6). The color of the biscuits showed that the two major protein fractions, the high-molecular-weight water solubles (H) and the salt-solution solubles (S), did not contain the pigments responsible for the color problem. The pigments responsible for the yellow coloration in biscuits were present in the low-molecular-weight water-soluble fraction (L). Pigments causing the brown color in biscuits were present in the water and salt-solution insoluble fraction (I).

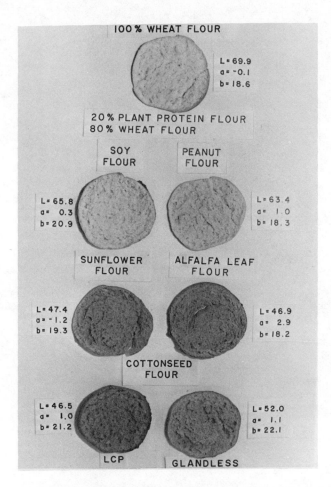

Figure 5. Biscuits containing 100% wheat and 20% plant-protein products

Journal of Food Science

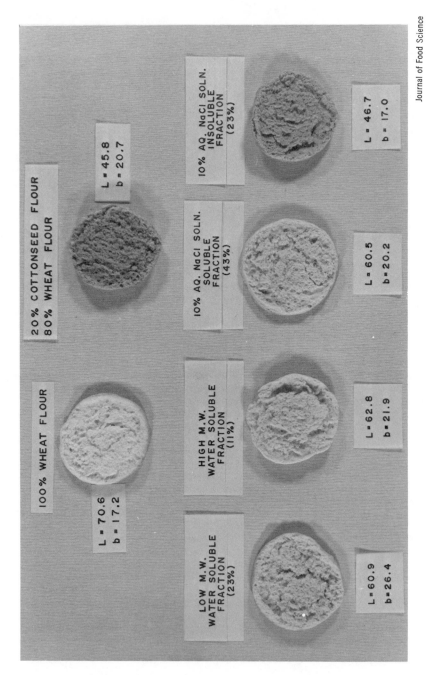

Figure 6. Biscuits containing 100% wheat flour, 20% LCP cottonseed flour, 4.5% L fraction, 2.5% H fraction, 8.5% S fraction, and 4.5% I fraction (4)

Role of Flavonoids

   The low-molecular-weight water-soluble fraction of LCP flour
was found by thin layer chromatographic methods to contain several flavonoid components. To establish the role of flavonoids
in the production of yellow color in biscuits, these components
were extracted from LCP and glandless cottonseed flours with 85%
aqueous isopropyl alcohol (which is a better solvent for flavonoids than water). Before removal of the flavonoids, the flours
had been treated with petroleum ether to extract residual lipids
that could interfere with flavonoid isolation. Extraction of
the residual lipids did not significantly alter the color of
biscuits prepared with the extracted flours (Figure 7).
Biscuits prepared with LCP and glandless cottonseed flours after
extraction with aqueous isopropyl alcohol to remove flavonoids
were brown; those prepared with the alcoholic extracts (in quantities equivalent to that present in the original flour) were
yellow (Figure 8).
   Thin layer chromatography (TLC) of the aqueous isopropyl
alcohol extracts showed that in addition to flavonoids considerable amounts of other components were in the extracts.
Consequently, the alcoholic extracts were next separated on a
Sephadex G-15 column, with water as the eluent, into nonflavonoid
(95%) and flavonoid (5%) fractions. Figure 9 shows two-dimensional TLC plates of the original extracts and the nonflavonoid
and flavonoid fractions of these extracts from both LCP and
glandless flours. The characteristic yellow flourescent spots
observed under ultraviolet light (long) indicate the presence of
at least six major flavonoids in the original extracts and the
flavonoid fractions. Because the concentrations used on the
plates for the original extracts and for the nonflavonoid fractions were the same, the absence of the yellow flourescent spots
for the nonflavonoid fractions indicates that the Sephadex column
fractionation was effective in separating the flavonoids from the
nonflavonoid components.
   Biscuits prepared with the nonflavonoid fractions of the
alcoholic extracts from LCP and glandless flours were tan and
near white, respectively (Figure 10). Biscuits containing the
flavonoid fractions were yellow. This very clearly establishes
that the yellow color in biscuits prepared with cottonseed flours
is caused by the flavonoids in these flours. This finding is
also supported by the observation that biscuits containing the
commercially available flavonoid, rutin, have the same yellow hue
and saturation as the biscuits containing the cottonseed flavonoids (Figure 11). Rutin is a flavonol glycoside with quercetin
as the aglycone and with a disaccharide of glucose and rhamnose
attached to the 3-hydroxyl position of the aglycone. It is, in
fact, one of the components found in the cottonseed flours.
Seven major flavonoids have been isolated from LCP flour and
tentatively identified (Table I).

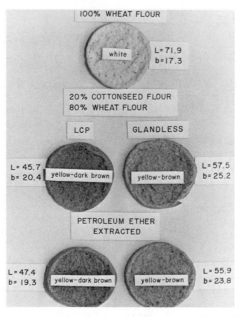

Figure 7. Biscuits containing 100% wheat flour, 20% LCP and glandless cotton-seed flour, and 20% LCP and glandless cottonseed flour after extraction with petro-leum ether (5)

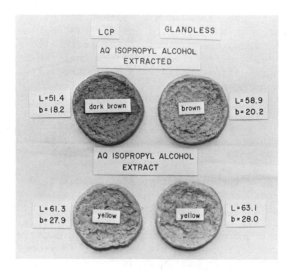

Figure 8. Biscuits containing 18.0% LCP and 18.5% glandless cottonseed flours after extraction with 85% aqueous isopropyl alcohol and extracts isolated from these flours, 1.8% and 1.5%, respectively (5)

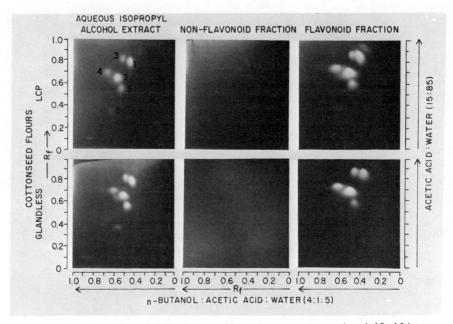

Figure 9. Two-dimensional cellulose TLC of flavonoids in aqueous isopropyl alcohol extracts, nonflavonoid fractions of extracts, and flavonoid fractions of extracts of LCP and glandless cottonseed flours (5)

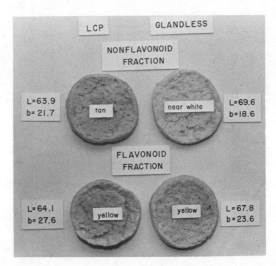

Figure 10. Biscuits containing the nonflavonoid (1.7% and 1.4%) and flavonoid (0.1% and 0.1%) fractions of the aqueous isopropyl alcohol extracts of LCP and glandless cottonseed flours, respectively (5)

Table I. Tentative Identification of Cottonseed Flavonoids

DESIGNATION[1]	TENTATIVE IDENTIFICATION	CHARACTERIZATION METHOD[4]
1	quercetin[2] 3-0-neohesperidoside (2-0-α-L-rhamnosyl-β-D-glucoside)	1,2,4
2	quercetin 3-0-glucoglucoside	1,2
5B	quercetin 3-0-robinoside (6-0-α-L-rhamnosyl-β-D-galactoside)	1,2
5A	quercetin 3-0-rutinoside or rutin (6-0-α-L-rhamnosyl-β-D-glucoside)	1,2,3
6	quercetin 3-0-glucoside or isoquercetrin (β-D-glucoside)	1,2,3
3	kaempferol[3] 3-0-neohesperidoside (2-0-α-L-rhamnosyl-β-D-glucoside)	1,2,4
4	kaempferol 3-0-glucoglucoside	1,2

1/ For number designations see Fig. 9.

2/ quercetin   3/ kaempferol

4/ (1) Ultraviolet – visible diagnostic spectral analysis.
   (2) Hydrolysis and TLC of aglycones and sugars.
   (3) Chromatographic mobility with standards.
   (4) NMR spectra.

The flavonoid mixture obtained by fractionation on the Sephadex G-15 column was further separated on a Sephadex LH-20 column, with 50% methanol as the eluent, into one fraction containing flavonoids 1, 3, and several minor flavonoids, and into a second fraction containing flavonoids 2 and 4 to 6 (see Figure 9 for number designations). These mixtures were then separated on thick layer cellulose plates into individual flavonoids. They were rechromatographed on thick layer plates until they appeared to be pure by two-dimensional TLC. As a final purification step, they were again chromatographed on a Sephadex LH-20 column. This final step yielded two components, designated A and B, from flavonoid 5. Diagnostic ultraviolet-visible spectral analysis indicated that flavonoids 1, 2, 5A, 5B, and 6 were 3-0-glycosides of quercetin and that flavonoids 3 and 4 were 3-0-glycosides of kaempferol. Acid hydrolysis of the flavonoids followed by TLC of the aglycones substantiated the above assignments. TLC of the hydrolyzed sugar components showed that flavonoids 2, 4, and 6 contained only glucose, flavonoids 1, 3, and 5A contained glucose and rhamnose and flavonoid 5B contained galactose and rhamnose. Chromatography with known flavonoids, rutin, and isoquercetrin, indicated the identity of flavonoids 5A and 6. The chemical shift of the H-1 rhamnosyl protons in the NMR spectra of flavonoids 1 and 3 indicated the disaccharides in these compounds are linked $\alpha-(1\rightarrow2)$ and are thus neophesperidosides. Flavonoid 5B because of its composition and chromatographic similarity with rutin is identified as a quercetin robinoside containing a rhamnosyl galactoside linked $\alpha-(1\rightarrow6)$. The linkage of the glucose disaccharide units in flavonoids 3 and 4 has not yet been established. The above identifications agree with the work of Pratt and Wender (12, 13) on cottonseed flavonoids.

The Role of Bound Gossypol Pigments

Gossypol is the major pigment in glanded varieties of cottonseed. It is a binaphthyl terpenoid and is toxic to monogastric animals (14). In the seed, this pigment is contained in structures referred to as pigment glands. The liquid cyclone process (1) separates these glands from the protein flour. Rupture of some of the glands during processing frees the gossypol. allowing it to react with the cottonseed proteins. A typical edible LCP flour contains less than 0.045% free gossypol and 0.26% bound gossypol (15).

One of the aldehyde groups of the gossypol is believed to react with the primary amine of the lysine units of the cottonseed proteins to form the Schiff base or imine type compound as shown at the top of Figure 12 (16, 17). An analytical method for determination of bound gossypol, developed by Pons et al. (18), involves treatment of the cottonseed flour with 3-amino-1-propanol and acetic acid in N, N-dimethylformamide (DMF)

Journal of Food Science

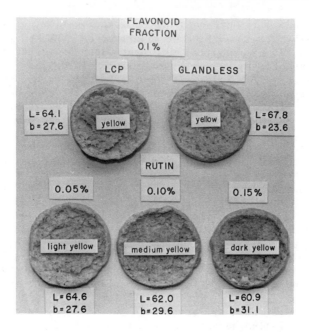

Figure 11. Biscuits containing flavonoid fractions of LCP and glandless cottonseed flours and rutin (5)

Journal of Food Science

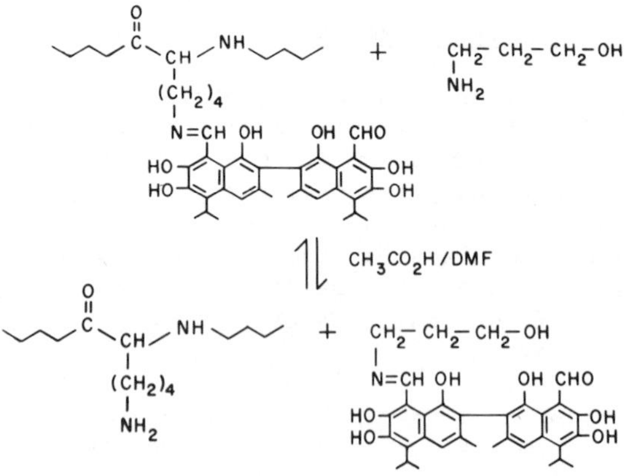

Figure 12. Reaction of protein-bound gossypol with 3-amino-1-propanol in the presence of acetic acid in DMF solution (4)

(Figure 12). Under these conditions, the gossypol and gossypol-like pigments become bound to the aminopropanol as the Schiff base of this low-molecular-weight compound. The gossypol-aminopropanol derivative can then be washed from the flour and measured spectrophotometrically.

This aminopropanol treatment for removal of bound gossypol appeared to be an excellent means to test the role of bound gossypol in the color problem. The water and salt-solution insoluble fraction (I) of LCP flour that contained the brown color causing pigments was treated for 15 min at 80°C with 2% 3-amino-1-propanol and 2% acetic acid in DMF solution. The aminopropanol extract was filtered off, and the flour residue was washed and dried. A biscuit prepared with the treated flour fraction showed that most of the brown color was removed by the aminopropanol treatment (Figure 13). This result suggests that bound gossypol pigments play a major role in the color problem in LCP flour.

In order to test this idea further, the extract of the aminopropanol treated flour was fractionated on a Sephadex LH-20 column with DMF as the eluent. The extract before fractionation gave an ultraviolet spectrum similar to that of standard gossypol treated with aminopropanol (Figure 14, curves B and A, respectively). Three colored fractions were obtained by column chromatography of the extract. The spectrum of the second band eluted (Figure 15, curve 2), was almost identical to the spectrum obtained from the gossypol-aminopropanol derivative (Figure 14, curve A). The first and last bands eluted from the column exhibited the spectra in curves 1 and 3 of Figure 15, which are similar to the gossypol-aminopropanol curve; however, they do differ considerably in the 350-450 nm region. These data indicate that bound gossypol and at least two bound gossypol-like pigments are responsible for the brown coloration in biscuits prepared with LCP cottonseed flour.

Other Color Components

Glandless cottonseed flour also caused yellow-brown color in biscuits, and a similar color distribution was obtained when biscuits were prepared from fractions of glandless flour (Figure 16). The intensity of the brown color was much less in the biscuit prepared with the water and salt-solution insoluble fraction of glandless flour than that prepared with the insoluble fraction of LCP flour (Figure 6). The brown color in this fraction of glandless flour was not affected by the aminopropanol treatment, which removed brown color from the LCP flour. Analysis of this flour indicated the absence of any detectable quantities of free or bound gossypol. The brown color in glandless flour is believed to be due to phenolic compounds of another type. The brown pigment containing fraction is composed mainly of insoluble

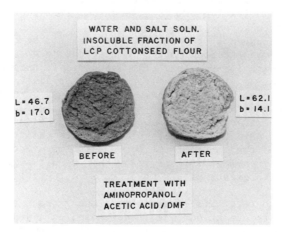

Figure 13. Biscuits containing the water and salt-solution insoluble fraction of LCP cottonseed flour before and after treatment with aminopropanol (4.5%)

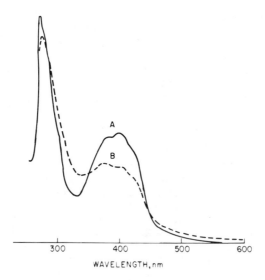

Figure 14. UV spectra of (A) gossypol-amino reaction mixture and (B) extract from aminopropanol treatment of water and salt-solution insoluble fraction of LCP cottonseed flour (4)

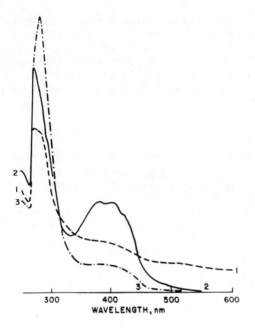

Figure 15.   UV spectra of Fraction 1, 2, and 3 separated from the aminopropanol extract of the water and salt-solution insoluble fraction of LCP cottonseed flour (4)

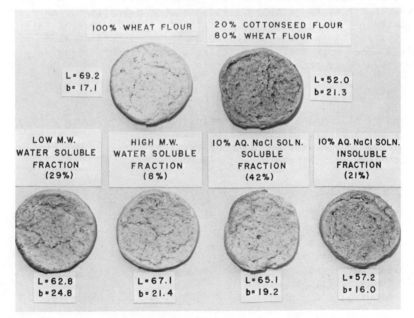

Figure 16.   Biscuits containing 100% wheat flour, 20% glandless cottonseed flour, 6.0% L fraction, 1.5% H fraction, 8.0% S fraction, and 4.0% I fraction

proteins and carbohydrates. If the remaining proteins are
removed by treatment with sodium dodecyl sulfate and
2-mercaptoethanol, most of the brown color remains with the
insoluble carbohydrate residue. These results suggest that the
brown color causing components in glandless cottonseed flour are
either bound to the polysaccharides (hemicellulose or cellulose)
or are insoluble polymers such as tannins or lignins.

Conclusions

When cottonseed flours and some other plant-protein products
are used in food applications, there is a serious discoloration
problem. Plant phenolic constituents are the major contributors
to this problem. When cottonseed flours are used in food products, a yellow coloration is caused by cottonseed flavonoids.
The seven major flavonoids in these flours were isolated and
identified as 3-0-glycosides of kaempferol and quercetin. The
marked brown discoloration observed when LCP glanded cottonseed
flour is used in a food product is caused by bound gossypol and
at least two bound gossypol-like pigments. The brown color
observed when glandless cottonseed flour is used in food is
believed to be due to other phenolic constituents that are
either insoluble polymers or are bound to the insoluble plant
polysaccharides.

Color in food products ranks second in importance to taste in
relation to consumer acceptability of a product. Discoloration
problems caused by plant-protein products must be solved if these
products are to be accepted. Isolation and identification of the
pigments producing color is an important step in solving this
problem and the methods developed in the studies presented in
this chapter with cottonseed flours are applicable to color problems caused by other plant-protein products.

Acknowledgment

Names of companies or commercial products are given solely for
the purpose of providing specific information; their mention does
not imply recommendation or endorsement by the U.S. Department of
Agriculture over others not mentioned.

Literature Cited

1. Gardner ᴴ. K., Jr.; Hron, R. J., Sr.; Vix, H. L. E. Cereal
    Chem., 1976, 53, 549.
2. Kadan, R. S.; Ziegler, G. M., Jr.; Spadaro, J. J. Cereal
    Chem., 1978, 55, 919.
3. Kim, M. K.; Calvin, B. M.; Lawhon, J. T. Cereal Sci.
    Today, 1971, 16, 216.
4. Blouin, F. A.; Cherry, J. P. J. Food Sci., 1980, 45, 953.

5. Blouin, F. A.; Zarins, Z. M.; Cherry, J. P. J. Food Sci., 1981, in press.
6. Mabry, T. J.; Markham, K. R.; Thomas, M. B. "The Systematic Identification of Flavonoids"; Springer-Verlag: N.Y., 1970; p. 354.
7. Judd, D. B.; Wyszecki, G., Eds. "Color in Business, Science, and Industry"; 2nd ed., Wiley and Sons, Inc.: N.Y., 1963; 500.
8. Francis, F. J.; Clydesdale, F. M., Eds. "Food Colorimetry: Theory and Applications"; Avi Publ. Co., Inc.: Westport, Conn., 1975; p. 477.
9. Sabir, M. A.; Sosulski, F. W.: Finlayson, A. J. J. Agric. Food Chem., 1974, 22, 575.
10. Sodini, G.; Canella, M. J. Agric. Food Chem., 1977, 25, 822.
11. Guilleux, F. Rev. Fr. Corps Gras, 1976, 23, 11.
12. Pratt, C.; Wender . H. J. Am. Oil Chem. Soc., 1959, 36, 392.
13. Pratt, C.; Wender, S. H. J. Am. Oil Chem. Soc., 1961, 38, 403.
14. Berardi, L. C.; Goldblatt, L. A. Gossypol, In: "Toxic Constituents of Plant Foodstuffs"; Ed. I. E. Liener; Academic Press: N.Y., 1969; pp. 211-266.
15. Gastrock, E. A.; D'Aquin, E. L., Eaves, P. H.; Cross, D. E. Cereal Sci. Today, 1969, 14, 8.
16. Baliga, B. P.; Lyman, C. M. J. Am. Oil Chem. Soc., 1957, 34, 21.
17. Conkerton, E. J.; Frampton, V. L. Arch. Biochem. Biophys., 1959, 81, 130.
18. Pons, W. A., Jr.; Pittman, R. A.; Hoffpauir, C. L. J. Am. Oil Chem. Soc. 1958, 35, 93.

RECEIVED September 5, 1980.

Flavor Volatiles as Measured by Rapid Instrumental Techniques

M. G. LEGENDRE

Southern Regional Research Center, Science and Education Administration, Agricultural Research, U.S. Department of Agriculture, P.O. Box 19687, New Orleans, LA 70179

H. P. DUPUY

V-Labs, Inc., 423 N. Theard St., Covington, LA 70433

Food flavor is governed by many factors, including lipid oxidation and protein degradation. Enzyme-catalyzed oxidation (1) and autoxidation (2) can substantially alter the flavor quality of foods. In addition, protein degradation, whether caused by enzymes, heat, or interactions with other compounds, can also affect flavor characteristics of certain foods (3, 4). Considerable effort has been made to examine the volatiles and trace components that contribute to food flavors. Some early techniques for measuring the volatile components in food products by gas chromatography consisted of analyzing headspace vapors to detect vegetable and fruit aromas (5) and volatiles associated with other food materials (6). Also, sample enrichment has been used in the analysis of some food products. However, these techniques require steam distillation or extraction and concentration, or both, before the volatile mixture can be introduced into a gas chromatograph (7, 8, 9, 10). Besides being complex and timeconsuming, these procedures often produce artifacts that can interfere with the analysis.

Dupuy and coworkers have reported a direct gas chromatographic procedure for the examination of volatiles in vegetable oils (11), raw peanuts and peanut butters (12, 13), and rice and corn products (14). When the procedure was applied to the analysis of flavor-scored samples, the instrumental data correlated well with sensory data (15, 16, 17), showing that food flavor can be measured by instrumental means. Our present report provides additional evidence that the direct gas chromatographic method, when coupled with mass spectrometry for the identification of the compounds, can supply valid information about the flavor quality of certain food products. Such information can then be used to understand the mechanisms that affect flavor quality.

Experimental Procedures

Materials. Tenax GC, 60-80 mesh (2,6-diphenyl-p-phenylene oxide) and Poly MPE (poly-m-phenoxylene) were obtained from

Applied Science Laboratories, State College, Pa. Teflon O-rings
were purchased from Alltek Associates, Arlington Heights, Ill.,
sandwich-type silicone septums from Hamilton Company, Reno, Nev.,
and Pyrex glass wool from Corning Glass Works, Corning, N.Y.
(The O-rings, septums, and glass wool were conditioned at 200 C
for 16 hr before use.) Inlet liners, 10 X 84 mm, were cut from
borosilicate glass tubing. The experimental food blend samples
were obtained from the Northern Regional Research Center, Peoria,
Ill. The rice samples came from the USDA rice experiment station
in Beaumont, Tex., and the peanuts from the USDA Coastal Plain
Experiment Station in Tifton, Ga.

Gas Chromatography (GC). A Tracor MT-220 gas chromatograph
with dual independent hydrogen-flame ionization detectors was
used in conjunction with a Westronics MT22 recorder and a Hewlett-
Packard 3354 laboratory automation system. The columns were
stainless-steel U-tubes, 1/8 in. OD, 10 ft long, packed with Tenax
GC that had been coated with 8% Poly MPE. Operating conditions
were: helium carrier gas, 35 ml/min in each column; hydrogen, 60
ml/min to each flame; air 470 ml/min (fuel and scavanger gas for
both flames). Inlet temperature was 160°C; detector was at 250°C.
Column oven was maintained at approximately 30°C (room temperature)
during the stripping period. After removal of the inlet liner,
the column was heated to 80°C within 2 min, then programmed 4°
C/min for 35 min. The final hold was at 220°C for about 30 min
until the column was clear. The Teflon O-ring was positioned at
the bottom of the GC inlet to provide a leak-proof seal. Elec-
trometer attenuation was 10 X 4.

Mass Spectrometry (MS). A Hewlett-Packard 5930A mass spectrometer was interfaced with a Tracor 222 GC. The ionization potential was 70 eV and the scanning limits were 33 to 450 amu. Scanning and data processing were accomplished with a FINNIGAN/INCOS
2300 mass spectrometer data system. A novel injection system (18)
was used in place of the standard injection port in conjunction
with the GC/MS system.

Sample Preparation and Analysis for GC. A 3-3/8 in. length
of 3/8 in. OD borosilicate glass tubing was packed with a plug of
volatile-free glass wool at the bottom, but 1/4 in. above the
bottom. The sample was weighed directly into the liner. Another
plug of glass wool was placed in the liner on top of the sample.
The septum nut, septum, and retainer nut of the GC were removed,
and the liner containing the sample was inserted into the inlet
of the GC on top of the Teflon O-ring. When the retainer nut was
tightened above the upper rim of the liner, a seal was formed
between the base of the inlet and the lower rim. After the inlet
was closed with the septum and septum nut, the carrier gas was
forced to flow through the sample. This assembly has been previously described (15).

Volatiles were rapidly eluted from the heated sample and transferred to the upper portion of the GC column, which was cooled to room temperature during the 20-min volatile-stripping period. The liner with the spent sample was removed, the inlet system was again closed, and the temperature of the oven was raised to 80°C. Temperature programming was then begun. When complete, the temperature was maintained on final hold until all volatiles were eluted. The oven was then cooled to room temperature for the next sample.

Sample Preparation and Analysis for MS. In a **typical analy**sis, conditions were the same as described for GC, except that the external injection system (18) was used to strip the volatiles from the sample. Briefly, the external injection system consists of three sections: the inlet assembly, the condenser assembly, and the six-port rotary valve. The sample is secured in the inlet liner, the sample liner is introduced into the inlet assembly, the six-port valve is switched to the "Inject" position, and the heater to the inlet is turned on to the desired temperature. As the sample is heated, the flow of carrier gas strips the volatiles from the sample. The volatiles then pass through the condenser assembly (which is cooled with compressed air to room temperature) through the valve and onto the cool GC column. After the stripping period, the six-port valve is put into the "Run/Purge" position (this isolates the sample from the column), and the column is temperature-programmed to resolve the volatiles. While this resolution is taking place, the sample liner is removed and the condenser is heated, venting the sample moisture to the atmosphere. This venting is done to render the sodium sulfate in the condenser to the anhydrous state. At the end of the run, both the GC oven and the condenser are cooled to room temperature for the next run.

Results and Discussion

The importance of direct gas chromatography and combined direct GC/MS to the food industry is demonstrated by the analysis of volatile flavor components and contaminants in experimental samples of rice, food blends, and raw and roasted peanuts. By examining these samples, we are able to investigate flavor systems that are probably associated with lipid oxidation, thermal degradation of protein, or protein interactions with other compounds.

For example, recently, while being processed, a new variety of rice produced an off-flavor, which was initially suspected to be germplasm related. However, when the volatiles in the control and experimental samples of the raw rice were examined by direct GC/MS, we found that the raw experimental rice contained significant amounts of methylbromide and dimethylsulfide, as shown in chromatogram B of Figure 1. Analysis of the processed experimental rice sample yielded a significant level of dimethylsulfide

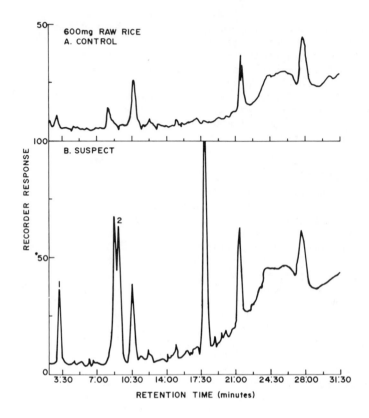

Figure 1.   Profiles of volatiles for control and experimental raw rice samples: (1)
methylbromide; (2) dimethylsulfide.

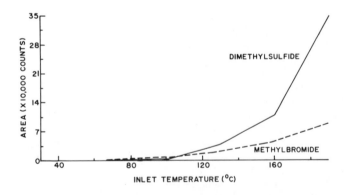

Figure 2.   Plot of area vs. inlet temperature for dimethylsulfide and methylbromide
from experimental raw rice samples

but no methylbromide. Neither compound was detected in the controls.

Methylbromide, which has been used in the food industry as a fumigant, has been postulated to react with methionine in peanuts and yield dimethylsulfide upon heating (19). These observations are also consistent with the findings of Bills et al. (21). These data were used to conclude that the suspected sample of experimental rice had been exposed to methylbromide, which caused the off-flavor. Thus, we have shown that the direct GC method can be used effectively to solve a practical problem.

To further support this conclusion, a series of raw samples of the experimental rice was subjected to varying temperatures between 25°C and 190°C in the external inlet of the gas chromatograph. As shown in Figure 2, the concentration of dimethylsulfide and methylbromide increased as the temperature of the inlet increased. The mass spectrometer was used to monitor the concentrations of methylbromide and dimethylsulfide. Masses 62 and 94 (the base peaks--largest ions in the mass spectra- of dimethylsulfide and methylbromide, respectively) were used to measure the concentrations of each compound. These findings are consistent with the results obtained by Bills et al. (21).

The direct gas chromatography/mass spectrometry method was also used to study the volatiles in corn-soy food blends without prior treatment or enrichment of the sample. As shown in chromatogram B of Figure 3, the inferior sample has a much greater concentration of volatile components than does the good (control) sample shown in chromatogram A. The inferior sample has large amounts of pentanal, 1-pentanol, hexanal, 1-hexanol, 2-heptanone, 2-pentylfuran, and the 2,4-decadienals compared to the good sample. These compounds are associated with lipid oxidation products. These results are analogous to those of Rayner et al. (22), who found that hexanal in soy protein products correlated well with taste panel flavor scores, and to those of Legendre et al. (14) with rice and corn products.

The direct GC/MS technique can be used to screen raw and processed materials. As shown in Figure 4, the volatiles in peanuts can be detected and identified both before and after roasting. When peanuts are exposed to roasting temperatures, many chemical processes take place. Among these processes is protein degradation, which is important in the development of good roasted peanut flavor and aroma. The formation of aldehydes, ketones, pyrroles, and pyrazines during roasting is shown in the lower chromatogram of Figure 4. The most intense peaks are the aldehydes and pyrazines. These compounds are produced by Strecker degradation of amino acids (20) and Browning reactions (23), respectively.

As mentioned above, this unique method can be used to detect and identify volatile contaminants in food materials. As shown in chromatogram A in Figure 5, raw peanuts that had been in a cardboard box in the laboratory yielded relatively large

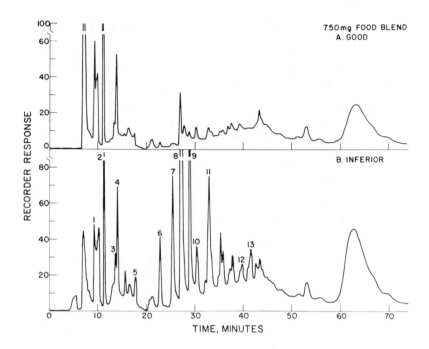

Figure 3.   Profiles of volatiles for good and inferior food blend samples: (1) acetal-
dehyde; (2) ethanol; (3) acetone; (4) methylene chloride; (5) chloroform; (6) pentanal; (7) 1-pentanol; (8) hexanal; (9) 1-hexanol; (10) 2-heptanone; (11) 2-pentyl-
furan; (12) trans-2,cis-4-decadienal; (13) trans-2,trans-4-decadienal.

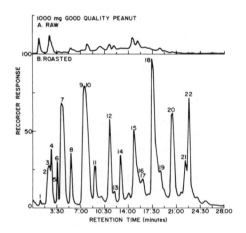

Figure 4.   Profiles of volatiles for high-quality peanuts, raw and roasted: (1) ethanol; (2) pentane; (3) 2-propanol; (4) acetone; (5) methylene chloride; (6) methyl
acetate; (7) 2-methylpropanal; (8) diacetyl; (9) 3-methylbutanal; (10) 2-methyl-
butanal; (11) 2,3-pentanedione; (12) N-methylpyrrole; (13) toluene; (14) hexanal;
(15) 2-methylpyrazine; (16) xylene; (17) 2-heptanone; (18) 2,5-dimethylpyrazine;
(19) 2-pentylfuran; (20) 2-ethyl-5-methylpyrazine; (21) 2-ethyl-3,6-dimethylpyra-
zine; (22) phenylacetaldehyde.

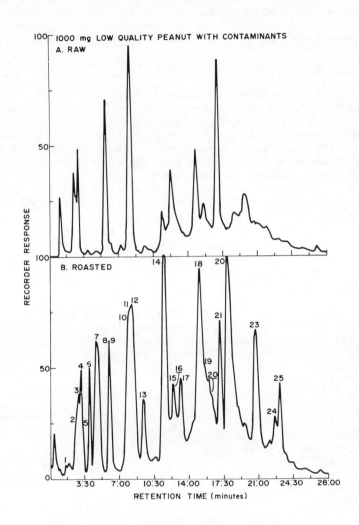

Figure 5.   Profiles of volatiles for low-quality peanuts with contaminants, raw and roasted: (1) ethanol; (2) pentane; (3) 2-propanol; (4) acetone; (5) methylene chloride; (6) methyl acetate; (7) 2-methylpropanal; (8) chloroform; (9) diacetyl; (10) benzene; (11) 3-methylbutanal; (12) 2-methylbutanal; (13) 2,3-pentanedione; (14) N-methyl-pyrrole; (15) pyridine; (16) toluene; (17) hexanal; (18) 2-methylpyrazine; (19) 2-methylpyrrole; (20) 2-heptanone; (21) styrene; (22) 2,5-dimethylpyrazine; (23) 2-ethyl-5-methylpyrazine; (24) 2-ethyl-3,6-dimethylpyrazine; (25) phenylacetalde-hyde.

amounts of methylene chloride, chloroform, benzene, pyridine, and styrene. As shown in chromatogram A of Figure 4, comparable samples of raw peanuts stored in a nonsolvent area did not yield these solvent materials. The concentration of these compounds decreased upon roasting, as shown in chromatogram B of Figure 5.

Conclusions

Rapid instrumental techniques were used to elucidate off-flavor problems in raw and processed rice products, raw and roasted peanuts, and corn-soy food blends. Less than a gram of the solid material was secured in a standard or special injection port liner of the gas chromatograph. Then, the volatiles from the sample were steam distilled in situ and identified by combined gas chromatography/mass spectrometry.

This unconventional approach provides a practical means of eluting, resolving, and identifying volatiles that might impart off-flavors (e.g., volatile components contributed by protein interactions, and external contaminants) in edible protein ingredients. It also provides information to resolve complex flavor problems for processors and plant breeders.

Acknowledgments

Names of companies or commercial products are given solely for the purpose of providing specific information; their mention does not imply recommendation or endorsement by the U.S. Department of Agriculture over others not mentioned.

Literature Cited

1. St. Angelo, A. J.; Legendre, M. G.; Dupuy, H. P. Lipids, 1980, 15, 45.
2. Chang, S. S. J. Am. Oil Chem. Soc., 1979, 56, 908A.
3. Minamiura, N.; Matsumura, Y.; Yamamoto, T. J. Biochem., 1972, 72, 841.
4. Mason, M. E.; Johnson, B. R.; Hamming, M. C. J. Agric. Food Chem., 1967, 15, 66.
5. Buttery, R. G.; Teranishi, R. Anal. Chem., 1961, 33, 1439.
6. Charalambous, G., Ed. in "Analysis of Foods and Beverages, Headspace Techniques"; Academic Press, Inc.: New York, 1978; p. xi.
7. Newar, W. W.; Fagerson, F. S. Food Technol., November, 1962, p. 107.
8. Chang, S. S. J. Am. Oil Chem. Soc., 1961, 38, 669.
9. Lee, C. H.; Swoboda, P. A. T. J. Sci. Food Agric., 1962, 13, 148.
10. Selke, E.; Moser, H. A.; Rohwedder, W. K. J. Am. Oil Chem. Soc., 1970, 47, 393.

11. Dupuy, H. P.; Fore, S. P.; Goldblatt, L. A. J. Am. Oil Chem. Soc., 1973, 50, 340.
12. Brown, M. L.; Wadsworth, J. I.; Dupuy, H. P.; Mozingo, R. W. Peanut Sci., 1977, 4, 54.
13. Fore, S. P.; Dupuy, H. P.; Wadsworth, J. I.; Goldblatt, L. A. J. Am. Peanut Res. Educ. Assoc., 1973, 5, 59.
14. Legendre, M. G.; Dupuy, H. P.; Ory, R. L.; McIlrath, W. O. J. Agric. Food Chem., 1978, 26, 1035.
15. Dupuy, H. P.; Rayner, E. T.; Wadsworth, J. I.; Legendre, M. G. J. Am. Oil Chem. Soc., 1977, 54, 445.
16. Williams, J. L.; Applewhite, T. H. J. Am. Oil Chem. Soc., 1977, 54, 461.
17. Jackson, H. W.; Giacherio, D. J. J. Am. Oil Chem. Soc., 1977, 54, 458.
18. Legendre, M. G.; Fisher, G. S.; Schuller, W. H.; Dupuy, H. P.; Rayner, E. T. J. Am. Oil Chem. Soc., 1979, 56, 552.
19. Cobb, W. Y.; Johnson, B. R. in " Peanuts--Culture and Uses"; American Peanut Research and Education Association, Stone Printing Co.: Roanoke, Virginia, 1973; p. 242.
20. Cobb, W. Y.; Johnson, B. R. in "Peanuts--Culture and Uses"; American Peanut Research and Education Association, Stone Printing Co.: Roanoke, Virginia, 1973; p. 254.
21. Bills, D. D.; Reddy, M. C.; Lindsay, R. C. Manuf. Confect., 1969, 49(9), 39.
22. Rayner, E. T.; Wadsworth, J. I.; Legendre, M. G.; Dupuy, H. P. J. Am. Oil Chem. Soc., 1978, 55, 454.
23. Newell, J. A.; Mason, M. E.; Matlock, R. S. J. Agric. Food Chem., 1967, 15, 767.

RECEIVED September 5, 1980.

Texturization

K. C. RHEE, C. K. KUO, and E. W. LUSAS

Food Protein Research and Development Center, Texas Engineering Experiment
Station, Texas A&M University System, College Station, TX 77843

Thermoplastic extrusion technology has been used to tex-
turize many defatted vegetable protein ingredients, and produce
many fibrous structures and meat-like textures. Such processes
have been used extensively to prepare meat analogs (1, 2) which
have found their widest application in formulation of foods for
institutional markets (3). A primary disclosure of extrusion
texturization of vegetable proteins was made by Atkinson (4).
General descriptions of various extrusion processes have also
been reported (1, 2, 5-14).

Ingredients most commonly used in textured vegetable protein
products are defatted soy flours or grits. Preferably, the defatted soy flour should contain a minimum of 50% protein with a
nitrogen solubility index of 50 to 70, a maximum of 30% insoluble
carbohydrate, and less than 1% fat (9).

While the technology of protein extrusion has been rapidly
developing, basic information concerning chemical and physical
changes occurring in components of raw ingredients within the
extruder barrel are still unknown. Most descriptions given are
general, and lack details. According to Harper (15), the process begins with defatted soy flour, which is moistened and
often mixed with a variety of additives. This mixture is fed to
the extruder where it is worked and heated, causing protein molecules to denature and form new cross-linkages which result in
a fibrous structure. The heated plasticized mass is forced
through a die at the extruder discharge to form expanded texturized strands of vegetable proteins which have meat-like
characteristics upon rehydration. However, the manner in which
the protein was denatured and cross-linked to form the fibrous
structure was not explained in this description.

Kinsella (16), in his recent review on texturized proteins,
described the texturization process as follows: the globular
proteins (glycinins) in the aleurone granules become hydrated
within the extruder barrel, are gradually unravelled, and are
stretched by the shearing action of the rotating screw flites.
The proteins become aligned in sheaths. In passing through the

0097-6156/81/0147-0051$09.50/0

die, the protein becomes further compressed and laminated longitudinally, and is denatured at high temperatures. At the instant of emergence from the die, pressure drops, moisture vaporizes and flashes off, and air-space vacuoles are produced within the laminated extrudate. The cooling, accompanied by vaporization, allows rapid thermosetting or solidification of the stretched protein fibers. A porous structure, with parallel arrays of lamellae of protein fibers, results. Again, how the protein is unravelled and the stretched protein is aligned in sheaths were not explained.

Most of the research conducted to date has been concerned with optimization of process conditions, and effect of their changes on selected physical characteristics and microstructure of the extrudate (17-22). Little information has been published concerning effects of ingredient compositions on extrudate characteristics. Proteins from different sources, varying in composition, molecular weight and structure, are expected to significantly affect the product upon texturization. For example, inclusion of gluten (prepared from hard wheat) in the feed mix yields a tough and chewy product, while use of gluten from soft wheat yields a tender and friable product (10). Reasons for these differences are still unknown.

The profound reason for having this many unknowns is attributed to the fact that the "technology" of extrusion texturization is well in advance of the "science". In-depth knowledge in the "science" portion of extrusion texturization will provide principles which can be used to accurately define the texturized protein product. This basic scientific knowledge should also provide data which could be used to improve operating efficiency of extruders, and serve as a basis for producing texturized protein foods, with desired characteristics for specific product applications, from a wide variety of raw ingredients.

EXPERIMENTAL PROCEDURES

Preparation of Raw Materials. Defatted soy flours used in this study were Soya Fluff-200W (Central Soya Inc., Fort Wayne, IN) and Extra Acted Soy Flour 200-I (A.E. Staley, Decatur, IL). These two flours met the requirements proposed by Smith (9).

Soy isolate was prepared by the isoelectric precipitation procedure developed at the Food Protein Research and Development Center, Texas A&M University System (23). A commercial soy isolate, Promine F (Central Soya Inc., Fort Wayne, IN), was also used in this study.

Modification of Raw Materials. In order to vary the protein and soluble sugar contents, portions of soy flour (Extra Acted Soy Flour 200-I) were replaced with various amounts of sucrose and soy protein isolate (Promine F). Levels of replacement were 5, 10 and 15% for sucrose, and 10, 20, 40 and 60% for soy isolate, respectively.

The pH of soy flour (pH 6.6) was altered by spraying water containing precalculated amounts of acid (HCl) or base (NaOH). When 1:3 (w/v) slurries were prepared from these modified flours, the actual pH values obtained were 5.3, 5.6, 6.6, 8.0 and 9.0, respectively.

To vary the Nitrogen Solubility Index (NSI) of the Soya Fluff 200W (NSI 65), 600-g batches of flour were heated on baking pans in an oven (135°C) for 2, 4 and 12 hr. These heat treatments reduced the 65% NSI flour to 51, 46 and 14% NSI, respectively. Extra Acted Soy Flour 200 I, with an NSI of 55.8%, was also included in this study without any heat treatment.

Ionic strength of the Soya Fluff 200W was varied by adding various amounts of NaCl and $CaCl_2$. Amounts of each salt added to the flour were 0.2, 1 and 2%, based on flour weight.

Molecular weights of soy flour proteins were enzymatically modified by the following method: (1) 20% soy flour slurry (w/v, pH 8)was shaken for 30 min in a 50°C water bath; (2) Papain (Wallerstein, Deerfield, IL) (1:1,000, w/v, papain to flour) was added to the slurry after shaking; (3) digestion was continued for 10 min at 50°C; (4) the digested slurry was heated rapidly to 85°C in a boiling water bath to stop enzyme activity; (5) after cooling, pH of slurry was adjusted back to 6.6 followed by freeze drying; and (6) dried product was then ground (100 mesh). A control was prepared in the same manner without the enzyme added.

In an attempt to determine the possible chemical reactions between free amino (NH_2) and carboxyl (COOH) groups during texturization, flour protein was chemically modified with various amounts of succinic and acetic anhydrides to block the selected number of free amino and carboxyl units to limit or inhibit reaction between these two groups. Soy flour was succinylated according to the method of Groninger (24). Fifty (50) grams of soy flour was dispersed in 250 ml of water, and various amounts (1.25, 2.50 and 3.75 g) of succinic anhydride were added to the dispersion at 0.5 g increments during constant stirring. The pH of the dispersion was maintained at 7.5 with 4N NaOH. After the pH had been stabilized, the dispersion was dialyzed against distilled water (4°C, 24 hr) to remove the excess succinic anhydride. The succinylated soy flour was then recovered by lyophilization. The acetylation procedures were similar to succinylation, except amounts of acetic anhydride added to each 50 g of soy flour were 2.5, 3.75 and 10 ml. Dialysis was continued for 72 hr with changing of distilled water every 24 hr.

To investigate the hypothesis that dissociation of disulfite bonds into subunits might be a reaction occurring during formation of texturized soy flour (25), two disulfide bond reducing agents (Na_2SO_3, and cysteine-HCl) were added to soy

flours at three concentrations (0.1, 0.2, 0.5% based on flour
weight) prior to extrusion. A strong oxidizing agent, KIO_3,
thought to be effective in reducing extensibility of wheat flour
dough by enhancing disulfide linkages, was also added to soy
flour at 0.01 and 0.05% concentrations.

Surfactants have been, reprotedly, used to prevent extensive
puffing of extruded cereal products. It was found in these
studies that surfactants could effectively inhibit gelatinization
of cereal starch. However, effect of surfactants on protein
texturization has not been reported. Two types of surfactants,
sodium stearoyl-2-lactylate and calcium stearoyl-2-lactylate (at
levels of 0.2 and 0.4% based on the weight of the flour), were
mixed with soy flour prior to extrusion. A yeast protein
(Torutein, manufactured by Amoco Inc.), claimed to be an
extrusion helper although its function is not known, was added.
In preliminary studies, it was found that this protein gives
results similar to a surfactant when added at 1% and 2% levels.

Texturization Processing. A Wenger Model X-5 laboratory
extruder was used in this investigation. This unit has a 1-inch
diameter extrusion barrel, with a barrel length to diameter ratio
of approximately 18:1. The barrel is composed of 8 jacketed
sections, and is fitted with a sidefeeder and a terminal
extrusion die plate. Diameter of the die orifice is 0.4 cm. The
screw is a TSP type originally designed and manufactured by
Wenger for extrusion of soy flour.

Conditions used for extrusion texturization of raw ingredients were: screw speed, 625 rpm; feed rate, 160 g/min; and
distance between end plate and screw end, 1/4 inch. The minimum barrel temperature to obtain a product surviving autoclaving
was around 138°C. Higher temperatures increased puffing of the
product. However, steam migration to the feed section caused
blockage. To minimize this blockage problem the first and
second sections of the extruder barrel were not heated. In order
to achieve a constant temperature of 138°C, steam pressures were
controlled at 2.7×10^4 kg/m^2. All raw materials were pre-
moistened to $25 \pm 0.5\%$ moisture, and equilibrated overnight
prior to extrusion.

The Hand Press (HP) texturizer described by Sterner and
Sterner (26) was used to prepare texturized products from
chemically modified soy flours. Although the Hand Press does
not produce shearing action during texturization, products prepared under proper conditions are morphologically and rheologi-
cally similar to extrusion texturized products (27). The Hand
Press texturizer consists of a heated circular recessed bottom
plate into which twenty to twenty-five grams of sample are
placed. A heated top plate is then placed over the sample, and
a controlled pressure applied to the top plate by a lever
mechanism, at a known temperature, for a predetermined length
of time (sec). The pressure is then rapidly released by removing

the hand from the lever. Rapid pressure release permits instantaneous vaporization of moisture to steam, which results in a textured wafer-shaped product.

In this Hand Press study all raw ingredients were processed under the conditions optimized for defatted soy flour: temperature (143°C-160°C); pressure, 250-300 psi; and retention time, 10 sec for 25-g samples with flour moisture contents of 25%.

Assay Methods for Evaluation of Flours and Texturized Products. Nitrogen solubility index of soy flour at neutral pH was determined by the Official and Tentative Methods of the American Association of Cereal Chemists (29).

Water holding capacity (WHC) of texturized products was determined by a relatively fast, easy and repeatable method. Extrudates were first dried to 5 to 7% moisture content, and reduced to 4-6 mesh size particles. To 5 grams of extrudate, 25 ml of water was added. After one hour hydration, the hydrated sample and residue water were filtered through No. 1 Whatman filter paper. The WHC was expressed as:

$$WHC = \frac{\text{Grams of water held}}{\text{Gram of dry sample}}$$

Water retention capacity (WRC), which is defined as grams of water retained per gram of dry extrudate after hydration and centrifugal filtration, was determined by the following procedure. One (10) gram of dry solid (particle size of 4-6 mesh) was weighed onto the sintered glass filter and hydrated for one hour with 10 ml water. A tared thimble, containing the sample and water, was then placed in a conical-shaped plastic centrifuge tube. The assembly was then spun for 15 min at 650 x g in an International No. 2 centrifuge. The WRC was then calculated as:

$$WRC = \frac{\text{Weight gain of sample}}{\text{Weight of dry sample}}$$

To determine bulk density (BD), extrudates or Hand Press Texturized products were dried to 5-7% moisture, and ground. Product passing through a U.S. No. 4 Sieve but retained on the U.S. No. 6 Sieve, was filled in 100 ml volumetric flasks, and tapped lightly for 20 times. Weight of the 100 ml sample was determined, and the bulk density was then expressed as grams per liter.

An Instron Testing System (Model 1122), fitted with a 10 cm^2 six-wire grid (Ottawa Texture measuring system, OTMS cell) was used to determine rheological properties. A loading rate of 50 mm/min and a chart speed of 500 mm/min resulted in a well defined force-deformation curve. Force at the bioyield point and the area under the curve were calculated. These values were then converted into maximum stress, work and specific work values:

stress (Newton (n)/cm^2)=force per unit area of test sample:Work
(N-cm/cm^3)=work per unit value of test sample; and specific work
(N-cm/g)=work per unit weight of test sample.

Samples for rheological evaluation by the Instron were
prepared by first reducing particle size to minus 4 plus 6
meshes, and rehydrating product with twice the amount of water
(w/w) for one hour in a covered glass beaker. The rehydrated
samples were then retorted at 121°C (15 psi) for 30 min. The
retorted samples were cooled in a running tap water bath, and
held in the room housing the Instron for 30 min before testing.
Five 7.5-g portions of samples were weighed onto the Ottawa cell
for the Instron measurement (30).

Protein solubilities of soy flour and extrudates in the
following solvent systems were determined by the micro-Kjeldahl
method (29). A portion (0.1 g) of finely-ground sample
(No. 60 sieve) was extracted with 9.9 ml of solvent for 1 hr at
room temperature followed by centrifugation and filtration. An
aliquot of the supernatant was used for nitrogen determination.
A factor of 6.25 was used to convert nitrogen content to protein.
Solvent systems used included: (1) 0.01M phosphate buffer, pH
7.2; (2) 0.01M carbonate buffer, pH 10.0; (3) Phosphate buffer +
1% 2-mercaptoethanol (2-ME); (4) Phosphate buffer + 1% sodium
dodecyl sulfate (SDS); and (5) Phosphate buffer + 1% 2-ME + 1% SDS.

The extent of succinylation and acetylation was determined
by the trinitrobenzenesulfonic acid (TNBS) method as described
by Hall et al. (31). Protein (5 mg) isolated from chemically
modified and control soy flour in 0.8 ml of aqueous solution was
added to 1 ml of 4% NaHCO$_3$, followed by addition of 0.2 ml of
2,4,6-trinitrobenzenesulfonic acid (12.5 mg/ml). The reaction
mixture was incubated at 40°C for 2 hr, and 3.5 ml of 36% HCl
was added. The tubes were stoppered and kept at 110°C for 4 hr.
After cooling to room temperature (24°C), volume of solution was
made up to 10 ml with distilled water, and contents were extracted twice with anhydrous diethyl ether.

The tubes were unstoppered and held at 40°C to allow the
residual ether to evaporate. The absorbance of the yellow (ε-
TNP lysine) solution was measured at 415 nm against a blank.
The extent of modification was calculated from the difference
in absorbance between the control and modified soy protein.

$$\% \text{ Modification} = \frac{A_{\text{unmodified}} - A_{\text{modified}}}{A_{\text{unmodified}}}$$

The sodium dodecylsulfate Polyacrylamide Gel Electrophoresis
(SDS-PAGE) was carried out according to the method of Weber and
Osborn (32). For the first 30 min of electrophoresis, a current
of 4 mA per gel (10 cm long) was used, and then increased to 8 mA
per gel for the subsequent 6 hr.

After electrophoresis, the dye front on the gel was marked
with ink, and the gel was stained with 2.5% Coomassie Brilliant

Blue solution in a mixture of methanol and glacial acetic acid
and water (5:1:5) for 2 hr. Finally, the stained gels were de-
stained in a solution containing 5% methanol and 7.5% acetic
acid in water using an automatic electric destainer.

For disc gel electrophoresis (DGE), a low current (1.5 mA
per gel) was applied to the gel during the initial period (when
the sample was still in the stacking gel) and the amperage was
increased to 3 mA per gel as soon as the samples penetrated into
the running gel. The rest of the procedure was the same as the
SDS-PAGE.

Preparations and Examinations of Samples for Transmitted
Light (TLM), Transmission Electron (TEM) and Scanning Electron
(SEM) Microscopies. Methods to prepare specimens for microscopic studies developed by Mollenhauer and Totten (33) and
modified by Cegla et al. (30) were followed. TLM examinations
were conducted on a Zeiss Standard 19 Research microscope. TEM
examinations were conducted on a Hitachi HS-8 at Kv on 600-800A°
sections prepared according to Galey and Nilsson (34). SEM
examinations were conducted on JOEL JSM-35 at 25 kv.

FACTORS AFFECTING EXTRUSION TEXTURIZATION PROPERTIES OF SOY FLOUR

Effects of Protein Contents. Protein contents of soy flours
were modified by adding various amounts of sucrose of soy protein
isolate (Promine F). Extrudates of sucrose-added soy flours
were not significantly different from the control in exterior
morphology. Scanning electron micrographs showed that sucrose-
containing extrudates were similar to the control in interior
structure (Figure 1). However, sucrose-containing extrudates
had lower stress and resilience values than the control when
measured with Instron, but they had higher bulk densities and
lower water retention and holding capacities than the control
(Table I).

Increasing the protein content of raw ingredients from
54.3% to 57.9% did not significantly change the interior and
exterior morphologies of extrudates. Physical and rheological
properties of these two products were also similar. However,
when the protein content was increased to 61.8% or higher, all
extrudates appeared larger in diameter, smoother in surface
morphology, lower in bulk density and higher in water retention
and holding capacities than the control.

Effects of Nitrogen Solubility Index. Nitrogen solubility
indexes of soy flour (initial NSI 65%) were modified by heating
in an oven (135°C) for 2, 4 and 12 hr prior to extrusion. These
treatments resulted soy flours with NSI values of 51, 46 and 14%,
respectively. The NSI-14 flour was found unusable for extrusion.
The original flour (NSI 65%) produced extrudates having low

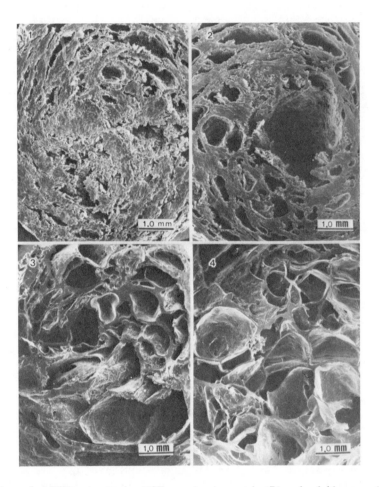

Figure 1.   SEMs of extrusion TSFs varying in protein (P) and soluble sugar (S) contents. (1) 57.9% P and 12.2% S, (2) 61.8% P and 6.2% S, (3) 69.3% P and 8.8% S, (4) 75.8% P and 6.2% S. Note that they are significantly different in air cell sizes.

TABLE I. Selected Physical and Rheological Properties of Extrusion Texturized Soy Products with Varied Protein and Soluble Sugar Contents[a]

Protein Content (%)	Soluble Sugar Content (%)	Bulk Density (g/l)	Water Retention Capacity (g/g)	Water Holding Capacity (g/g)	Stress (N/cm²)	Resilience (N-cm/cm³)	Specific Work (N-cm/g)	Loss of Integrity after Retorting[b]
45.9	27.1	446.0a	0.63f	2.05g	2.82f	1.09e	1.89e	+
48.6	24.6	446.0a	0.70e	2.10f	3.04ef	1.10e	1.90e	+
51.3	20.4	435.1b	0.72e	2.20e	3.54e	1.13e	1.96e	-
54.2	14.2	411.3c	0.85d	2.18e	5.28d	1.88d	3.26d	-
57.9	12.4	418.5c	0.95c	2.37d	6.82c	2.42c	4.19c	-
61.8	11.4	355.3d	0.98c	2.56c	7.94b	3.12b	5.41b	-
69.3	8.8	300.1e	1.07b	2.93b	8.40b	3.67a	6.36a	-
75.8	6.2	238.2f	1.27a	3.32a	10.90a	3.48a	6.03a	-

[a] Means with different letter in the same column differ significantly at P<0.05.

[b] "+" signs mean loss of integrity after retorting.

bulk density, but high water retention and holding capacities
(Table II).

Rheological properties of extrudates were not significantly
related to NSI values of raw ingredients. However, these extrudates appeared quite different on transmitted light micrographs
and exterior morphology. Higher NSI extrudates showed more
fibrous and continuous protein-carbohydrate matrices. Smoothness
and continuity of surface, and diameter of extrudates, decreased
as flour NSI decreased. Higher extruder feed rates (240 g/min)
could also change physical and rheological properties of low NSI
extrudates, but did not affect the exterior morphology.

Effects of pH. pH of soy flour was modified by adding
various amounts of 1N HCl or 1N NaOH. Experimental data indicate
that physical properties of extrudates were related to pH values
of raw ingredients; bulk density was negatively correlated (r =
-0.85) and water retention and holding capacities were positively
correlated (r = 0.91 and 0.89, respectively) (Table III). Extrudates with lower pH values had stronger texture, as expressed
in terms of Instron stress and resilience values. Also, pH
values of raw ingredients greatly affected exterior morphology
and interior structure of extrudates. Surface smoothness and extrudate diameters increased as flour pH became alkaline (Figure
2). Fibrousness of protein matrix in extrudates decreased as
flours became acidic in pH (Figure 3).

Effects of Molecular Weight Distributions. Effects of
molecular weight distributions of soy proteins on morphology,
microstructure, and physical and rheological properties of nonextrusion (Hand Press) texturized soy products had been previously
reported (35). Enzyme-modified soy flours, which contained 69%
low molecular weight polypeptides (<10,000 g/mole), could not
form retort-surviving products. According to Burgess and Stanley
(36), energy produced in the extruder is high enough to cause
free amino and carboxyl groups to form peptide bonds. If true,
enzyme-modified soy flour should be able to form a product which
is similar to a normal texturized soy product.

In this study, a Wenger X-5 extruder was used to process
enzyme-modified soy flours to determine effects of molecular
weight distributions on textural properties. Extruded products,
produced from enzyme-modified flour (TEMSF), showed significant
differences from unmodified products (TSF) in exterior morphology
(Figure 4) and also in microstructures as determined by scanning
electron and transmitted light micrographs (Figure 5). TEMSF
was much smaller in diameter and had smoother surfaces than TSF.
Examination of interior structures of extrudates revealed that
TSF had small air cells with thick air cell walls and rough air
cell surface. On the other hand, most of the air cells in TEMSF
were large with thin walls and smooth surfaces. Under the transmitted light microscope, TSF showed continuous protein fibers

TABLE II. Selected Physical and Rheological Properties of Extrudates as Affected by Nitrogen Solubility Index of Raw Ingredients[a]

Nitrogen Solubility Index	Bulk Density (g/l)	Water Retention Capacity (g/g)	Water Holding Capacity (g/g)	Stress (N/cm²)	Resilience (N-cm/cm³)	Specific Work (N-cm/g)	Loss of Integrity after Retorting[b]
65	459.0e	1.02a	2.46b	5.90c	2.25e	3.90	−
55	481.4d	0.97b	2.48b	5.28d	1.88d	3.26	−
51[c]	588.6a	0.75c	2.10e	5.42d	1.87d	3.24	+
51[d]	459.6e	0.99ab	2.54a	6.28b	2.23c	3.86	−
46[c]	577.2b	0.70d	2.18d	5.86c	2.45b	4.25	+
46[d]	493.8c	0.98b	2.40c	6.61a	2.69a	4.66	−

[a] Means with different letter in the same column differ significantly at $P < 0.05$.

[b] "+" signs mean loss of integrity after retorting.

[c] Extruded at low feeding rate (160 g/min).

[d] Extruded at high feeding rate (240 g/min).

TABLE III. Selected Physical and Rheological Properties of Extrudates as Affected by pH Values of Raw Ingredients[a]

pH	Bulk Density (g/l)	Water Retention Capacity (g/g)	Water Holding Capacity (g/g)	Stress (N/cm^2)	Resilience (N-cm/cm^3)	Specific Work (N-cm/g)	Loss of Integrity after Retorting[b]
5.3[c]	574.8a	0.58e	1.94c	6.70b	2.46b	4.18d	+
5.3[d]	334.3e	0.83d	2.44b	8.90a	3.37c	5.84a	-
5.6	563.2a	0.61e	2.01c	6.82b	2.35b	4.05b	+
6.6	459.0b	1.02c	2.46b	5.90c	2.25b	3.90c	-
8.0	419.0c	1.07b	2.66b	4.87d	1.81c	3.14d	-
9.0	379.8d	1.16a	2.89a	4.98d	1.75c	3.03e	-

a/ Means with different letter in the same column differ significantly at P<0.05.

b/ "+" signs mean loss of integrity after retorting.

c/ Extruded at low feeding rate (160 g/min).

d/ Extruded at high feeding rate (240 g/min).

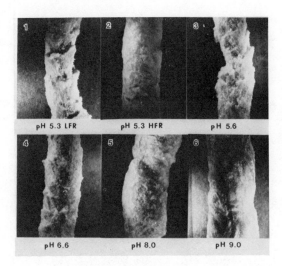

Figure 2. Exterior morphologies of soy extrudates varying in pH value and feeding rate. LFR: (160 g/min); HFR: (240 g/min). Note the increasing diameter and surface smoothness when the pH value of soy flour increased. Higher feeding rate could change the characteristics of low-pH-value extrudates. (Mag. 1.8×.)

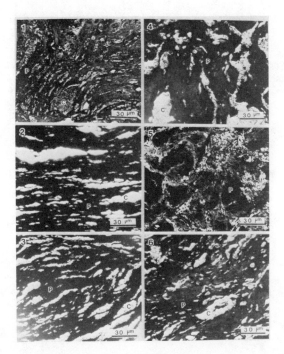

Figure 3. TLMs of extrusion texturized pH-modified soy flours: (1) pH 9.0; (2) pH 8.0; (3) pH 6.6; (4) pH 5.6; (5) pH 5.3 extruded at LFR; (6) pH 5.3 extruded at HFR. Note that alkaline pH could increase the fibrousness of the protein matrix; acidic pH produced the opposite effect. P, protein; C, insoluble carbohydrate.

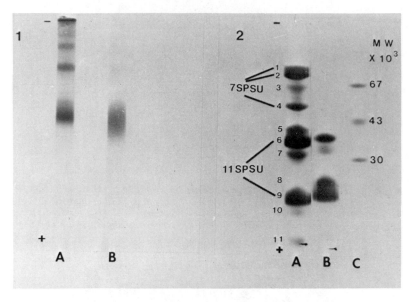

Figure 4.   PAGE patterns of unmodified (A) and enzyme-modified (B) soy proteins
with protein markers (C) in the absence (1) or presence (2) of SDS and 2-mercapto-
ethanol. 7S PSU, 7S protein subunits; 11S PSU, 11S protein subunits.

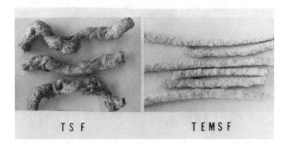

Figure 5.   Photographs of extrusion TSF and TEMSF. Note the differences in
exterior morphology and diameter.

with evenly dispersed cell wall fragments, whereas TEMSF showed discontinuous protein-carbohydrate matrix (Figure 6). The most significant difference between these two products was the texture. TEMSF lost its textural integrity upon hydration with cold water while TSF retained textural integrity even after retorting (Table IV).

Apparently, the energy level attained in the extruder is not high enough to form stable linkages among smaller molecules produced by enzymatic hydrolysis of soy proteins (Figure 7). High molecular weight proteins (>50,000 g/mole) are needed to form texture during extrusion of soy protein ingredients.

Effects of Ionic Strengths. Two types of salts (NaCl and $CaCl_2$), at three concentrations (0.2, 1.0 and 2.0%), were added to soy flours prior to extrusion to change the ionic strength. It appeared that effects of NaCl were quite different from those of $CaCl_2$. Neither salt affected the texture of extrudates at low concentrations (0.2%); however, at higher concentrations, NaCl increased the bulk density but hindered texture formation, while $CaCl_2$ decreased bulk density but increased textural strength of extrudates. NaCl-added extrudates could not survive the retorting process, while $CaCl_2$-added extrudates could (Table V). NaCl affected surface smoothness and diameter of extrudates, whereas $CaCl_2$ affected only the diameter. Addition of NaCl decreased the degree of fibrousness of protein matrix in transmitted light micrographs; however $CaCl_2$ showed opposite effects.

Effects of Surfactants. Two types of surfactants [sodium stearoyl lactylate (SSL) and calcium stearoyl lactylate (CSL)] at two concentrations (0.2 and 0.4%) were added to soy flours to study the effect of surfactants on the texturization properties of soy flours. An extrusion helper (Tolutein, yeast cell protein) was also included at two concentrations (1 and 2%) in this study because of its behavior similar to a surfactant upon extrusion.

All surfactant-added extrudates showed higher bulk densities, and lower water retention and holding capacities than the control (Table VI). Effects of SSL and CSL on physical properties of extrudates were similar. The extent of Tolutein effects at 1 and 2% levels on physical properties of extrudates was similar to that of SSL and CSL at 0.2 and .4% levels, respectively. Addition of surfactants significantly reduced the Instron stress and resilience values of extrudates. The most pronounced difference between control and surfactant-added extrudates was the ability to withstand the retorting process. All extrudates surfactants containing became mushy after autoclaving. They showed a rough discontinuous surface morphology (Figure 8) and less fibrous protein matrix (Figure 9) than the control.

TABLE IV. Selected Physical and Rheological Properties of Extrusion Texturized Unmodified and Enzyme-modified Soy Flours

	Bulk Density (g/l)	Water Retention Capacity (g/g)	Water Holding Capacity (g/g)	Stress (N/cm^2)	Resilience (N-cm/cm^3)	Specific Work (N-cm/g)	Loss of Integrity after Retorting [a]
TSF[b]	372.4	1.21	2.70	7.67	2.00	3.46	–
TEMSF[c]	238.9	n.d. [d]	n.d.	n.d.	n.d.	n.d.	+

[a] "+" sign means loss of integrity after retorting.

[b] Texturized soy flour.

[c] Texturized enzyme-modified soy flour.

[d] Values could bot be calculated due to the product did not retain the structural integrity after rehydration.

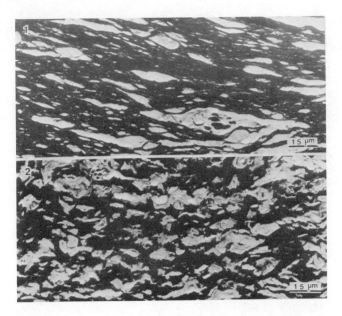

Figure 6. TLMs of (1) extrusion TSF and (2) extrusion TEMSF. The lighter areas are insoluble carbohydrates or cell wall fragments, and the darker areas are proteins.

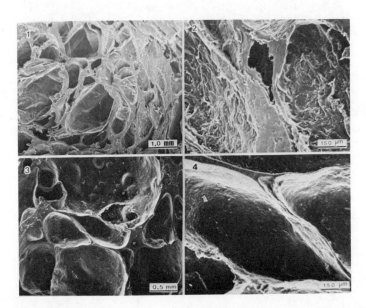

Figure 7. SEMs of (1) general morphology and (2) detailed air cell surface of TSF, and (3) general morphology and (4) detailed air cell surface of extrusion TEMSF

TABLE V. Selected Physical and Rheological Properties of Extrusion Texturized Soy Products as Affected by Different Types and Amounts of Salts[a/]

Salt	Bulk Density (g/l)	Water Retention Capacity (g/g)	Water Holding Capacity (g/g)	Stress (N/cm²)	Resilience (N-cm/cm³)	Specific Work (N-cm/g)	Loss of Integrity after Retorting[b/]
0.2% NaCl	467.8c	0.98b	2.48b	6.18d	2.10d	3.67cd	-
1.0% NaCl	497.9b	0.85de	2.32c	5.48d	2.03d	3.52e	+
2.0% NaCl	534.6a	0.82e	2.36c	4.76f	1.36e	2.36f	+
0.2% CaCl₂	452.3c	0.92c	2.52ab	7.00c	2.58ab	4.47b	-
1.0% CaCl₂	401.7d	1.00ab	2.57a	8.37b	2.43b	4.21b	-
2.0% CaCl₂	466.8c	0.81e	2.53a	10.06a	3.02a	5.23a	-
Control	459.0c	1.02a	2.46b	5.90e	2.25c	3.90c	-

[a/] Means with different letter in the same column differ significantly at $P < 0.05$.

[b/] "+" signs mean loss of integrity after retorting.

TABLE VI. Selected Physical and Rheological Properties of Extrusion Texturized Soy Products as Affected by Different Types and Amounts of Surfactants[a]

Surfactant	Bulk Density (g/l)	Water Retention Capacity (g/g)	Water Holding Capacity (g/g)	Stress (N/cm²)	Resilience (N-cm/cm³)	Specific Work (N-cm/g)	Loss of Integrity after Retorting[b]
1.0% Tolutein[c]	513.0e	0.85b	2.34b	5.78b	2.17a	3.76a	+
2.0% Tolutein	568.9c	0.82b	2.22c	4.70c	1.49d	2.58d	+
0.2% CSL[d]	525.7d	0.82b	2.24c	5.56b	1.78c	3.08c	+
0.6% CSL	575.5b	0.75c	2.13d	4.40c	1.27e	2.20e	+
0.2% SSL[e]	569.6c	0.81b	2.13d	6.25a	2.04b	3.53b	+
0.4% SSL	589.0a	0.73c	2.11d	5.50b	1.74c	3.01c	+
Control	459.0f	1.02a	2.46a	----	----	----	-

a/ Means with different letter in the same column differ significantly at P<0.05.

b/ "+" signs mean loss of integrity after retorting.

c/ Brand name of Amoco yeast protein.

d/ Calcium stearoyl lactylate.

e/ Sodium stearoyl lactylate.

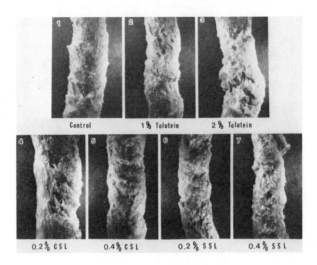

Figure 8. Exterior morphologies of extrusion texturized surfactant-added soy flours. Control, no surfactant added product; Tolutein, yeast protein; CSL; SSL. Note the effect of surfactant on the diameter and surface smoothness of extrudates. (Mag. 1.5×.)

Figure 9. TLMs of extrusion texturized surfactant-added soy flours: (1) 0.2% SSL added; (2) 0.4% SSL added; (3) 0.2% CSL added; (4) 0.4% CSL added; (5) 1% Tolutein added; (6) 2% Tolutein added. Note that all surfactants could decrease the fibrousness of the protein matrix. This effect became more significant at higher surfactant concentrations. P, protein; C, insoluble carbohydrate.

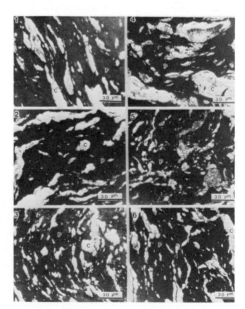

POSSIBLE ROLES OF CARBOHYDRATES IN TEXTURIZATION OF SOY FLOUR

Scanning electron and transmitted light microscopic techniques were used to study the roles of soluble and insoluble carbohydrates in texture formation during nonextrusion texturization processing of soy flours. The role of soluble carbohydrates was studied by comparing micrographs of texturized soy flour, texturized water extracted soy flour, and texturized soy concentrate. The role of insoluble carbohydrates was studied by comparing micrographs of texturized soy flour, texturized soy concentrate and texturized soy isolate with soy hull blends of these three products. Possible interactions between proteins and soluble sugars were investigated by comparing solubilities of proteins and sugars before and after texturization.

Possible Roles of Soluble Carbohydrates in Texture Formation. Microscopic structure of texturized water-extracted soy flour and texturized soy concentrate were quite similar to that of texturized soy flour. Scanning electron microgrpahs showed that water extraction of soy flours had little effect on morphological characteristics of texturized soy products (Figure 10). Solubility of soluble sugars was not affected by texturization, whereas solubility of proteins decreased sharply when soy flour was texturized (Table VII). It appears that soluble sugars did not interact with proteins during texturization. Based upon results of microscopy and solubility studies, it is reasonable to speculate that natural soluble carbohydrates are not required (do not play an important role) in development of texture or stabilization of structure.

Possible Roles of Insoluble Carbohydrates in Texture Development. Based on data obtained from soy isolate-soy hull blend texturization experiments, insoluble carbohydrates and crude fiber play an important role in modulating the morphology of final texturized products (Figures 11-14). Insoluble carbohydrates, because of their plastic response to deformation, control the type of alveolation developed during processing. They also appeared to control the type of cuticle morphology exhibited in the product's alveoli.

Only small amounts of insoluble carbohydrates were decomposed during texturization at higher temperatures. These decomposition products were not identified.

PHYSICAL AND CHEMICAL CHANGES OF SOY PROTEINS DURING TEXTURIZATION PROCESS

Morphological Changes of Soy Flours Extruded at Various Temperatures. Soy flour (Soyafluff 200W) was extruded at temperatures from 79 to 149°C, and extrudates were examined visually and microscopically. Extrudates produced at temperatures up to

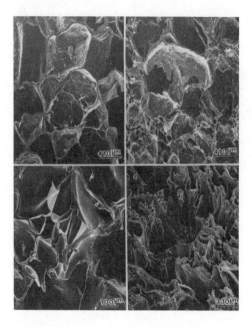

*Figure 10.   SEMs comparing air cell size and thickness of nonextrusion TSF (1),
soy concentrate (2), soy isolate (3), and aqueous-extracted soy flour (4)*

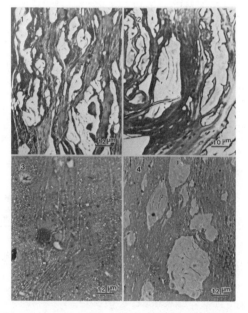

*Figure 11.   TLMs comparing microscopic structure of nonextrusion TSF (1), soy
concentrate (2), soy isolate (3), and aqueous-extracted soy flour (4)*

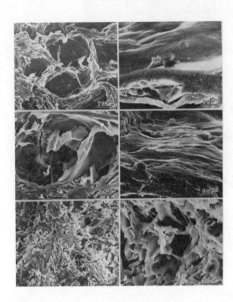

Figure 12. SEMs of nonextrusion TSF with different amounts of crude fiber (CF): (1) alveolate morphology of 5.3% CF product; (2) air cell cuticle of 5.3% CF product; (3) alveolate morphology of 9.3% CF product; (4) air cell cuticle of 9.3% CF product; (5) alveolate morphology of 13% CF product; and (6) air cell cuticle of 13% CF product.

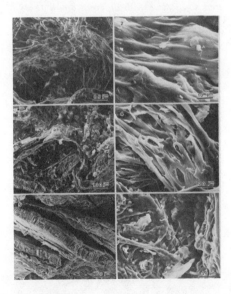

Figure 13. SEMs of nonextrusion textured soy concentrate with different amounts of CF: (1) alveolate morphology and (2) air cell cuticle of 5.7% CF products; (3) alveolate morphology and (4) air cell cuticle of 10.2% CF products; and (5) alveolate morphology and (6) air cell cuticle of 13.7% CF products.

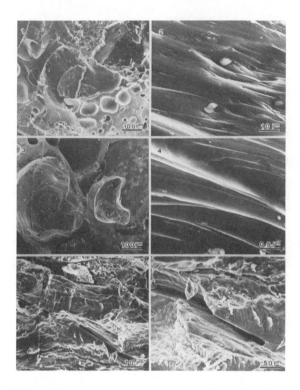

Figure 14.   SEMs of nonextrusion textured soy isolate with different amounts of
CF: (1) alveolate morphology and (2) air cell cuticle of 5.3% CF products; (3)
alveolate morphology and (4) air cell cuticle of 9.6% CF products; and (5) alveolate
morphology and (6) air cell cuticle of 13.1% CF products.

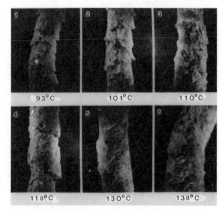

Figure 15.   Exterior morphologies of
soy extrudates produced at different temperatures.   Note that the diameter and
surface smoothness increased when the
extrusion temperature increased. (Mag.
1.2×.)

Table VII. Change in Amounts of Soluble Sugar and Protein
in Nonextrusion Texturized Soy Flour Processed
at Different Temperatures

Processing Temp. (°C)	Protein solubility[a] (%)	Soluble sugar[b] (g/100g)
no heat treatment	52.59	14.2
79	22.82	14.7
93	16.31	14.4
107	14.53	14.2
116	12.60	14.3
124	11.86	14.6
132	9.38	14.3
141	9.59	14.1
149	8.68	14.8
160	9.00	14.8

[a] Means of 2 measurements.

[b] Means of 4 measurements.

118°C did not puff when the products left the extruder die
(Figure 15). These extrudates were recognized as "undertex-
turzed". The common appearance of these undertexturized
products was lack of smooth and continuous surface. A minimum
temperature of 138°C was needed to produce puffed and retort-
resistant extrudates when other process conditions were held
constant.

Transmitted light micrographs (Figure 16) showed that, at
low temperatures (80 and 93°C), proteins aggregated into a
number of large discontinuous chunky masses, some of which
contain entrapped cell wall fragments. As the extrusion
temperature was increased, the protein-carbohydrate aggregates
became elongated fibrous matrix, but was still lacking continuity. When temperature was further increased to 138°C, most
of the water in the flour reached the varporization point, and
large amounts of proteins became plasticized, and the material
leaving the extruder die was expanded into a sponge-like structure. Examination of the interior structure of these extrudates
revealed proteins in a fibrous matrix, in which broken-down
cell wall fragments were dispersed evenly throughout.

Ultrastructure Changes in Soy Flours Induced by Nonextrusion
Texturization Process at Various Temperatures. Changes in soy
flour components during nonextrusion texturization at various
temperatures were unlike those of extrusion texturization because
shearing action does not occur in nonextrusion texturization.
Therefore, a higher temperature (160°C) was required to produce
a puffed product. Due to lack of shearing action, nonextrusion

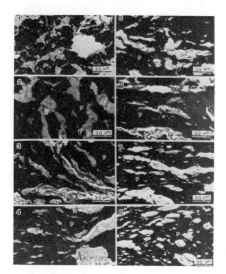

Figure 16. TLMs of soy extrudates produced at different temperatures: (1) 80°C; (2) 93°C; (3) 101°C; (4) 110°C; (5) 118°C; (6) 130°C; (7) 138°C; and (8) 145°C. Note that protein fiber formation is temperature dependent. A continuous fibrous matrix forms at about 130°C.

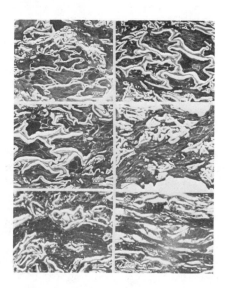

Figure 17. TLMs of TSFs processed with Hand Press at different temperatures ((1) 80°C; (2) 93°C; (3) 101°C; (4) 110°C; (5) 118°C; (6) 130°C; (7) 138°C; and (8) 145°C) showing sequential changes of cell and protein body, and how the fibrous matrix is formed.

texturization did not disrupt cells at 80°C (Figure 17). When temperature was increased, protein bodies became disrputed, and cell walls opened. When the protein bodies were further denatured, they began unfolding and fused with each other. At the same time, cell wall fragments were forced together to form small colonies. At 138°C, most of the proteins were fused into an expandable mass. When the temperature reached 149°C, most of the water became vaporized. With a sudden release in pressure, the compressed soy flour puffed and formed a plexilamellar matrix.

Chemical Reactions Occurring in Proteins During Extrusion at Various Temperatures. Possible chemical reactions occurring in proteins during extrusion, thought to be responsible for insolubilization of proteins during texturization, were studied using five buffer systems of well-defined functions. Various buffer-soluble proteins were further subjected to sodium dodecyl sulfate (SDS)-polyacrylamide gel electrohphoresis (PAGE) to determine extent of protein fractions insolubilized at given temperatures, and the nature of chemical reactions. By combining solubility data (Figure 18) with SDS-PAGE results (Figures 19-20), the pH 7.2 buffer-soluble proteins in extruded products were found to consist mostly of low molecular weight 11S protein subunits. Almost all 7S and high molecular weight 11S protein subunits were insolubilized. In 1% SDS-phosphate buffer, portions of the low molecular weight 7S and high molecular weight 11S protein subunits were solubilized. In 1% 2-mercaptoethanol (2-ME) buffers, the soluble proteins consisted mostly of high molecular weight 11S protein subunits. The remaining protein fractions could be solubilized in 1% SDS- or 1% 2-ME-containing buffers. These results strongly indicate that noncovalent forces (hydrogen bond, ionic bond and hydrophobic force) and sulfhydryl-disulfide interchanges are the major chemical reactions occurring during extrusion texturization.

Texturization Properties of Succinylated Soy Flours. To investigate mechanisms of intermolecular peptide bond formation responsible for texture formation and protein insolubility upon texturization, soy flours were chemically modified with various amounts of succinic anhydride to succinylate different amounts of free amino groups and, consequently, control the extent of intermolecular pepetide bond formations. Modified soy flours were nonextrusion texturized, and compared with the unmodified product for microscopic structure (Figures 21-23) and physical, chemical and rheological properties (Table VIII). All succinylated products were significantly different from unmodified products in microscopic structure and physical, chemical and rheological properties. Due to the fact that succinylation not only prevents coupling reactions between free amino and carboxyl groups, but also produces additional negatively

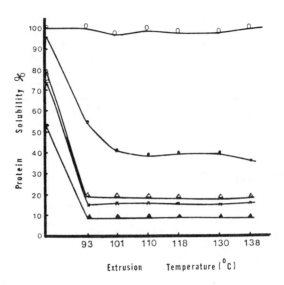

Figure 18.   Protein solubilities in various buffer systems of soy extrudates produced at different temperatures: (▲) pH 7.2, 0.0M phosphate buffer; (✕) pH 10.0, 0.01M carbonate buffer; (△) 1% 2-mE added phosphate buffer; (●) 1% SDS added phosphate buffer; (○) 1% 2-mE and 1% 2-mE added phosphate buffer.

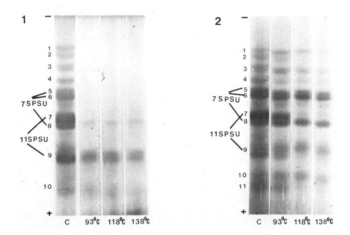

Figure 19.   SDS–PAGE patterns (without 2-mE) of various temperature-extruded soy proteins soluble in pH 7.2, 0.01M phosphate buffer without (1) or with (2) 1% SDS. 7S PSU, 7S protein subunits; 11S PSU, 11S protein subunits; C, control (unheated soy protein).

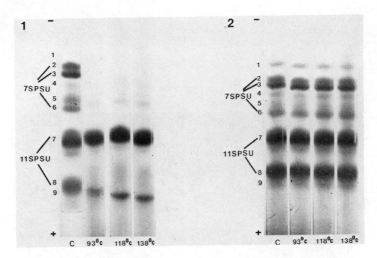

Figure 20. SDS–PAGE patterns of various temperature-extruded soy proteins soluble in (1) 1% 2-mE added phosphate buffer and (2) 1% 2-mE and 1% SDS added phosphate buffer. 7S PSU, 7S protein subunits; 11S PSU, 11S protein subunits; C, control (unheated soy protein).

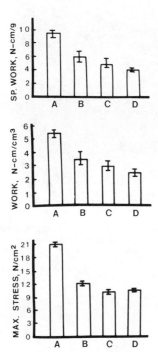

Figure 21. Rheological properties of nonextrusion textured (A) soy flour, (B) 25.5% succinylated soy flour, (C) 61.7% succinylated soy flour, and (D) 83% succinylated soy flour. The bars, indicating standard error, that do not overlap indicate a significant difference between the respective means at P = 0.05.

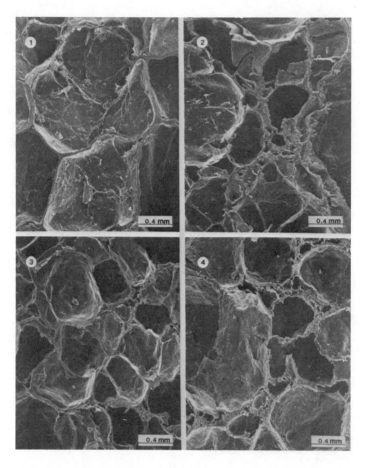

Figure 22.   SEMs showing air cell size and air cell wall thickness of textured (1) soy flour, (2) 25.5% succinylated soy flour, (3) 61.7% succinylated soy flour, and (4) 83% succinylated soy flour

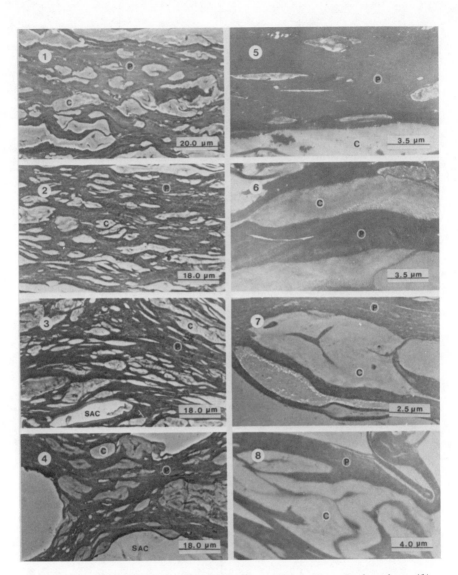

Figure 23. TLMs and TEMs showing the ultrastructure of texturized products: (1) TLM of texturized soy flour; (2) TLM of texturized 25.5% succinylated soy flour; (3) TLM of texturized 61.7% succinylated soy flour; (4) TLM of texturized 83% succinylated soy flour; (5) TEM of texturized soy flour; (6) TEM of texturized 25.5% succinylated soy flour; (7) TEM of texturized 61.7% succinylated soy flour; and (8) TEM of texturized 83% succinylated soy flour. SAC, small air cell; C, insoluble carbohydrate; and P, protein.

TABLE VIII. Protein Solubility and Selected Physical Properties of Extrusion Texturized Succinylated Soy Flours[a]

Succinic Anhydride[b]	Degree of Succinylation (%)	Protein Content (%)	Protein Solubility in Phosphate Buffer, pH 7.2 (%)		Water Retention Capacity (g/g)	Bulk Density (g/l)
			Before Texturization	After Texturization		
0	0	62.0	61.3	7.4	1.92c	135a
1.25	25.5 ± 5.2	61.9	85.5	11.9	3.02b	122b
2.50	61.7 ± 2.9	60.8	87.5	19.0	4.10a	116c
3.75	83.0 ± 2.9	59.1	91.6	23.8	4.05a	112d

a/ Means with different letter in the same column differ significantly at P<0.05.
b/ Amount of succinic anhydride, in grams, added to 50 g flour.

charged groups, succinylation might have exerted additional influences on texturization. Therefore, differences observed could have been due to reduced degrees of coupling reaction or repelling forces produced by highly negatively charged molecules, or the combined effects of these two reactions. However, although succinylation had two-fold effects on texturization, highly succinylated soy flours could still produce texturized products which could withstand retorting process without losing textural integrity. This result might suggest that intermolecular peptide bonds may not play an important role in texture formation.

Texturization of Disulfide Bond Reduced and Oxidized Soy Flours. To further assure that sulfhydryl-disulfide interchange is a possible mechanism for texture and structure formation of texturized soy products, soy flours were treated with two types of disulfide reducing agents (Na_2SO_3 and cysteine-HCl) at two concentrations. Soy flours were also treated with disulfide bond enhancing oxidizing agent (KIO_3) at two concentrations. Results of nonextrusion texturization of these treated flours revealed that soy flours could be texturized at lower temperatures in the presence of reducing agents (Figure 24). This could mean that cleavage of disulfide bonds an essential chemical reaction during texturization. Results of the extrusion texturization were similar to those of nonextrusion texturization studies. When other process conditions were held constant, addition of reducing agents produced extrudates having lower bulk densities and higher water holding capacities than the control (Table IX).

Addition of oxidizing agent to soy flour resulted in a less-puffed or undertexturized extrudates. The extrudate showed rough, discontinuous surface and small diameter. This result further proves that the change of disulfide linkages is an essential reaction for texture formation.

Transmitted light micrographs showed that all extrudates containing reducing agents had highly fibrillated protein-carbohydrate matrices which are similar to the control, whereas the extrudates containing oxidizing agents showed less fibrous structure (Figure 25).

CONCLUSIONS

The term "texture" can be broadly defined as "the composite of properties which arise from structural elements, and the manner in which it is perceived by the physiological senses." This definition recognizes three essential elements of texture: that it is a sensory quality; that it stems from structural parameters of foods; and that it is a composite of several properties. Considerable efforts have been made to understand more clearly the complex nature of texture and texture-forming

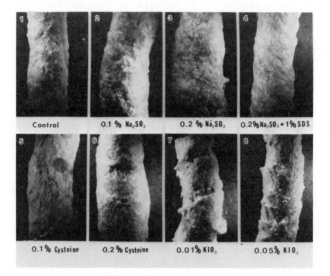

Figure 24. Exterior morphologies of extrusion texturized reducing or oxidizing agent-modified soy flours. Note that both reducing agents (Na₂SO₃ and cysteine) -modified flours produced more puffed products, whereas the oidant (KIO₃) modified flour showed opposite effect. (Mag. 1.5.)

TABLE IX. Selected Physical and Rheological Properties of Extrusion Texturized Soy Products as Affected by Added Reducing and Oxidizing Agents[a]

Additive	Bulk Density (g/l)	Water Retention Capacity (g/g)	Water Holding Capacity (g/g)	Stress (N/cm^2)	Resilience (N-cm/cm^3)	Specific Work (N-cm/g)	Loss of Integrity after[b] Retorting
Control	459.0c	1.02b	2.46c	5.90d	2.25c	3.90c	−
0.1% Cysteine-HCl	432.5d	0.34c	2.24ef	6.53bcd	2.46bc	4.26c	−
0.2% Cysteine-HCl	369.2f	1.06b	2.34d	7.54b	2.64b	4.58b	−
0.1% Na$_2$SO$_3$	423.9e	1.05b	2.26e	6.20cd	2.36c	4.09c	−
0.2% Na$_2$SO$_3$	316.1g	1.04b	2.73b	7.26bc	2.72b	4.64b	−
0.01% KIO$_3$	473.4b	0.90c	2.20f	3.88e	1.49d	2.58d	+
0.05% KIO$_3$	496.4a	0.74d	2.16g	3.64e	1.48d	2.56d	+
0.2% Na$_2$SO$_3$ +1% SDS[c]	150.0h	1.23a	3.96a	8.76a	2.99a	5.18a	−

a/ Means with different letter in the same column differ significantly at $p < 0.05$.

b/ "+" signs mean loss of integrity after retorting.

c/ SDS: sodium dodecyl sulfate.

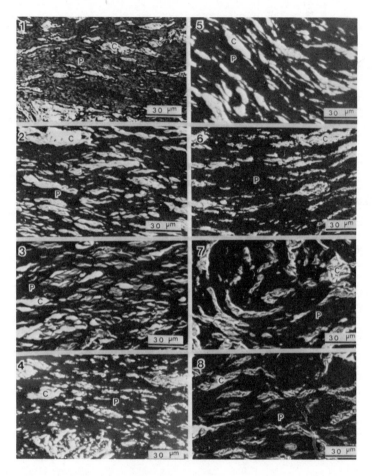

Figure 25.   TLMs of extrusion texturized chemically modified soy flours: (1) control; (2) 0.1% Na₂SO₃ added; (3) 0.2% Na₂SO₃ added; (4) 0.1% cysteine–HCl added; (5) 0.2% cysteine–HCl added; (6) 0.2% Na₂SO₃ and 1% SDS added; (7) 0.01% KIO₃ added; and (8) 0.05% KIO₃ added. Note that both Na₂SO₃ and cysteine–HCl can increase the fibrousness of protein matrix but KIO₃ has the opposite effect. P, protein; C, insoluble carbohydrate.

processes, by identifying the essential physico-chemical and functional properties, and structural requirements of ingredients for texture formation; studying the reaction mechanisms and interactions between components of ingredients during texturization; developing processes and equipment, and optimizing process conditions; developing methodologies for determination of various textural properties using instruments; and finally, by correlating the objective instrumental data with that obtained through sensory evaluation. As a result of concerted efforts, it has become known that: proteins with certain molecular sizes and solubilities are essential for texture formation; insoluble carbohydrates play significant roles; process temperature and pressure are important; and certain instrumental data can be correlated with the sensory evaluation data with reasonable accuracy and reproducibility. Additionally, some of the key reactions taking place during texturization of plant protein ingredients, as well as their possible mechanisms have been elucidated; however, large unknown areas still exist where future research activities should be directed to put the "art" of texturization on a sound scientific basis.

LITERATURE CITED

1. Horan, F.E. J. Amer. Oil Chem. Soc. 51:67A (1974).
2. Horan, F.E. Meat Analogs in "New Protein Foods, Vol. 1", Altschul, A.M., ed., Academic Press, New York, NY (1974).
3. Robin, R.F. Food Technol. 26:59 (1972).
4. Atkinson, W.T. U.S. Pat. 3,488,770 (1970).
5. Gordon, A. Food Process. Mark. 38:267 (1969).
6. Horan, F.E. and Wolf, H. Meat Analogs — A supplement in "New Protein Foods", Vol. 2, Altschul, A.M., ed., Academic Press, New York, NY (1975).
7. Rakosky, J. J. Agric. Food Chem. 18:1005 (1970).
8. Smith, O.B. Food Eng. 47:48 (1975).
9. Smith, O.B. Textures by Extrusion Processing in "Fabricated Food" Inglett, G.E., ed., AVI Publishing Co., Westport, CT. (1975).
10. Smith, O.B. Extrusion Cooking in "New Protein Foods", Vol. 2B, Altschul, A.M., ed., Academic Press, New York, NY (1976).
11. Wilding, M.D. J. Amer. Oil Chem Soc. 48:489 (1971).
12. Williams, M.A. Cereal Foods World 22:152 (1977).
13. Williams, M.A., Horn, R.E. and Rugala, F.P. Food Eng. 49:87 (1977).
14. Ziemba, J.V. Food Eng. 41(11):72 (1969).
15. Harper, J.M. CRC Critical Rev. Food Sci. Nutr. Feb. pp. 155 (1979).
16. Kinsella, J.E. CRC Critical Rev. Food Sci. Nutr. Nov. pp. 147 (1978)

17. Aguilera, J.M., Kosikowski, F.W. and Hood, L.R. J. Food
     Sci. 41:1209 (1976).
18. Cumming, D.B., Stanley, D.W. and deMan, J.M. J. Inst.
     Can. Sci. Technol. Aliment 5:124 (1972).
19. Maurice, T.J., Burgess, L.D. and Stanley, D.W. Can.
     Inst. Food Sci. Technol. J. 9:173 (1976).
20. Taranto, M.V., Meinke, W.W., Cater C.M. and Mattil, K.F.
     J. Food Sci. 40:1264 (1975).
21. Taranto, M.V., Cegla, G.F. and Rhee, K.C. J. Food Sci.
     43:767 (1978).
22. Taranto, M.V., Cegla G.F. and Rhee, K.C. J. Food Sci.
     43:973 (1978).
23. Circle, S.J. and Smith, A.K. Processing soy flours,
     protein concentrates and isolates in "Soybean: Chemistry
     and Technology", Vol. 1, Protein, Smith A.K. and Circle,
     S.J., ed., AVI Publishing Co., Westport, CT (1972).
24. Groninger, H. J. Agric. Food Chem. 21:978 (1973).
25. Cumming, D.B., Stanley, D.W. and deMan, J.M. J. Food Sci.
     38:320 (1973).
26. Sterner, M. and Sterner, H. Food Eng. 3:77 (1976).
27. Taranto, M.V. and Rhee, K.C. J. Food Sci. 43:1274 (1978).
28. AOCS. Official and Tentative Methods of the American Oil
     Chemists' Society, 3rd ed., Chicago, IL (1971).
29. AACC. Approved Methods of the American Association of
     Cereal Chemists, Vol. II, St. Paul, MN (1973).
30. Cegla, G.F., Taranto, M.V., Bell, K.R. and Rhee, K.C. J.
     Food Sci. 43:775 (1978).
31. Hall, R.J., Teinder, N. and Givens, D.I. Analyst 98:673
     (1973).
32. Weber, K. and Osborn, M. Proteins and sodium dodecyl
     sulfate: Molecular weight determination on polyacrylamide
     gel and related procedures in "The Protein", Vol. 1,
     Academic Press, New York, NY (1975).
33. Millenhauser, H.H. and Tolten, C. J. Cell Biol. 48:387
     (1971).
34. Galey, F.R. and Nilsson, S.E.G. J. Ultrastruct. Res. 14:
     405 (1966).
35. Kuo, C.M., Taranto, M.V. and Rhee, K.C. J. Food Sci. 43:
     1848 (1978).
36. Burgess, L.D. and Stanley, D.W. Can. Inst. Food Sci.
     Technol. 9:228 (1976).

RECEIVED November 11, 1980.

5

Solubility and Viscosity

JEROME L. SHEN

Ralston Purina, 900 Checkerboard Square, St. Louis, MO 63188

The manner in which proteins behave in a given food system or food application, i.e. its functionality, is a manifestation of the fundamental physicochemical properties of the proteins under the given conditions. Food systems are generally very complex, involving water, several types of protein as meat or soy, fats, salts, flavor and color compounds. Further, the functional properties of the food systems, generally, and that of soy proteins, particularly, are sensitive to past processing history, methods of preparation, and conditions of measurement. Because of this, it has been difficult to standardize functional testing, generalize the findings, and use the measured results from one system to predict the behavior in another system. Measurements that can have general applicability must be based upon the fundamental physiochemical properties of the components. Further, in order to predict functional behavior of the protein, it is necessary to determine the particular physicochemical states and interactions of the protein that result in the desired functionality.

Solubility and viscosity are two experimentally measurable properties that can yield information about the functional behavior as well as the physicochemical nature of the proteins. In this chapter, the application of these two measurements to isolated soy proteins and how these measurements can explain the basic physicochemical nature of isolated soy proteins are discussed.

Solubility

The requirements for the thermodynamic definition of solubility are 1) well defined initial solid and final solution states, and 2) establishment of equilibrium between these two states. Under these conditions, the solubility at a given temperature and pressure is the concentration of the sample in solution. At a given temperature and pressure, the solid state is the stable crystalline state; and the solution state is

0097–6156/81/0147–0089$05.25/0
© 1981 American Chemical Society

fixed with regard to such variables as degree of protein association, dissociation and protein conformation. Defined in
this manner, the thermodynamic solubility is independent of
path. As long as the initial and final states are unchanged,
the solubility will not depend upon how the final state is
achieved. For example, in the following reaction

A (25°C, 1Atm) $\rightleftarrows$ A (25°C, 1Atm, pH 7, I=.1M)
    Crystal   Solution

it makes no difference whether the final equilibrium is
approached by dissolving more and more crystalline A until the
solution is saturated, by starting with a saturated solution of
A at 35°C and cooling to 25°C, or by dissolving A in water
and then adding salt until 0.1M ionic strength is reached. As
long as equilibrium is established, the solubility will be
unchanged. This type of behavior when plotted in percent
soluble protein versus amount of solid protein added is
illustrated in Figure 1. As solid A is added to the solvent,
100 percent of the added protein will remain in solution until
a saturation limit is reached. After that, the percent in
solution will drop off and approach zero asymptotically as the
amount of added protein approaches infinity.

In contrast, Figure 2 shows that the percentage of protein
in solution for soy isolates remains constant as the amount of
added protein is increased (1). In other words, the amount of
protein in solution increases linearly with increasing amounts
of added protein. This behavior is observed for all the isolates we have studied up to the highest concentration of 18
percent. Thus, soy isolates behave as if they are composed of
a completely soluble fraction (A) and a completely insoluble
fraction (B). Upon the addition of solvent, the soluble fraction (A) dissolves completely while the insoluble fraction (B)
remains unchanged. There is no equilibrium established between
A and B such that, if B is separated from A and reslurried in
additional amounts of solvent, no additional protein will go
into solution. (More precisely, no evidence of microscopic
reversibility was found on the time scale of the experiment,
i.e. 2 hrs. at 25°C. Actually, for all the isolates studied,
the amount of protein in solution at 25°C reaches a maximum
plateau after 100 minutes in a shaker bath. Thus, at least
steady-state conditions are reached in our experiments. It is
not known whether much longer equilibration times, days, will
change the amount of soluble protein.) Thus, a key requirement
for thermodynamic solubility is not met.

Further, neither the initial solid state nor the final
solution state are well-defined. The initial state is a
heterogeneous amorphous solid. The distribution of the
fractions A and B depends upon the previous history of the
sample, such as manner of extraction, precipitation, solvent

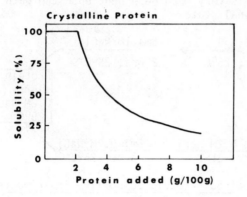

Figure 1. Solubility behavior of a pure crystalline protein

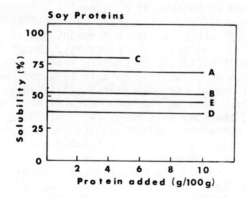

Figure 2. Solubility behavior of soy protein isolates (1)

treatment, heat treatment, and drying. The amount and the
nature of the protein in solution depends upon how the sample
is put into solution (slurry and blending methods, Figure 3),
upon how the insoluble fraction is removed (centrifugation
conditions, Figure 4), and upon the path by which the final
state is reached. For example, in the following experiment,
the solubility is very much dependent upon the path for
reaching the final state.

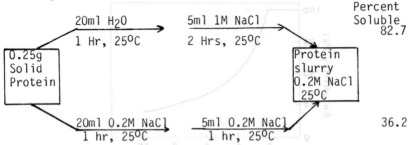

		Percent Soluble
20ml H_2O, 1 Hr, 25°C	5ml 1M NaCl, 2 Hrs, 25°C	82.7
0.25g Solid Protein	Protein slurry 0.2M NaCl 25°C	
20ml 0.2M NaCl 1 hr, 25°C	5ml 0.2M NaCl 1 hr, 25°C	36.2

Interpretation of Soy Protein Solubility Experiments

This observed behavior of soy proteins complicates the
definition of soy protein solubility, the comparison of solubility data, and the interpretation of solubility experiments.
Because the thermodynamic criteria are not met, protein solubility becomes an operationally defined quantity that depends
upon the experimental methods of measurement. A number of
different operational definitions have been used to measure
protein solubility. Each has its own advantages and disadvan-
tages, and limited utility. This plurality, though often
desirable, makes it difficult to compare experimental results.
Interpretation of solubility data in terms of the forces
and the interactions at the molecular level is an even more
difficult problem. Because thermodynamic criteria are not met,
straightforward, thermodynamic analyses cannot be applied.
Further, strict comparison of the results for soy proteins with
the results of systems that meet the thermodynamic criteria
cannot be justified. However, it may be valid to draw some
qualitative insights into the nature of soy protein by making
comparisons under favorable circumstances. One such favorable
case would be the following hypothetical mechanism

$$\text{Reaction I} \qquad nA \rightleftarrows A_n$$
$$\text{Reaction II} \qquad A_n \longrightarrow B$$

in which soluble monomers A reversibly form aggregates (A_n)
followed by the irreversible conversion of the aggregate (A_n)
into the insoluble precipitate (B). The aggregates are like
the insoluble precipitates in all respects except for the
factors that prevent B from being reversibly resolubilized. If

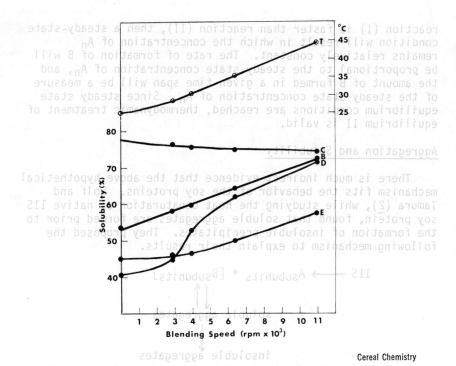

Cereal Chemistry

Figure 3.   The effect of blending on solubility (1). Samples A, B, D, and E are soy protein isolates. Sample C is a commercial sodium caseinate. T is the temperature of the slurry after blending.

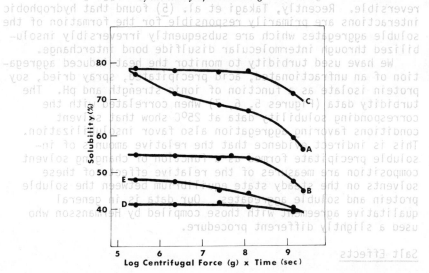

Cereal Chemistry

Figure 4.   The effect of centrifugation on solubility (1). Samples A, B, D, and E are soy protein isolates. Sample C is a commercial sodium caseinate.

reaction (I) is faster than reaction (II), then a steady-state condition will result in which the concentration of A_n remains relatively constant. The rate of formation of B will be proportional to the steady state concentration of A_n, and the amount of B formed in a given time span will be a measure of the steady state concentration of A_n. Since steady state equilibrium conditions are reached, thermodynamic treatment of equilibrium II is valid.

Aggregation and Solubility

There is much indirect evidence that the above hypothetical mechanism fits the behavior of the soy proteins. Wolf and Tamura (2), while studying the heat denaturation of native 11S soy protein, found that soluble aggregates are formed prior to the formation of insoluble precipitates. They proposed the following mechanism to explain their results.

$$11S \longrightarrow A_{subunits} + [B_{subunits}]$$

$$\updownarrow$$

soluble aggregates

$$\downarrow$$

insoluble aggregates

Catsimpoolas et al. (3) and Hermansson (4) have found that the reaction leading to formation of the soluble aggregates is reversible. Recently, Takagi et al. (5) found that hydrophobic interactions are primarily responsible for the formation of the soluble aggregates which are subsequently irreversibly insolubilized through intermolecular disulfide bond interchange.

We have used turbidity to monitor the heat induced aggregation of an unfractionated, acid precipitated, spray dried, soy protein isolate as a function of ionic strength and pH. The turbidity data (Figures 5, 6, 7) when correlated with the corresponding solubility data at 25°C show that solvent conditions favoring aggregation also favor insolubilization. This is indirect evidence that the relative amounts of insoluble precipitate formed as a function of changing solvent composition are measures of the relative effects of these solvents on the steady state equilibrium between the soluble protein and soluble aggregates. Our data is in general qualitative agreement with those compiled by Hermansson who used a slightly different procedure.

Salt Effects

Neutral salts are known to exert striking effects on the solubility, the association-disassociation equilibrium, the

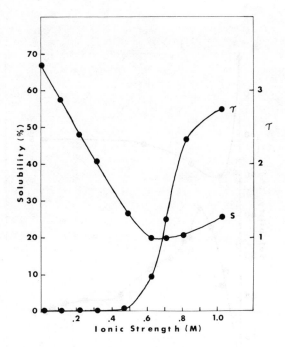

Figure 5. The effect of ionic strength on solubility (S) and the heat-induced turbidity (τ) at pH 2

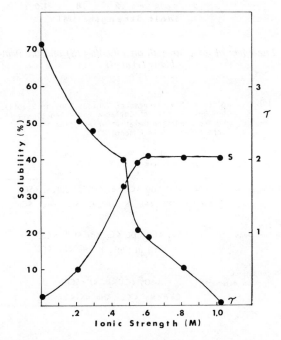

Figure 6. The effect of ionic strength on solubility (S) and the heat-induced turbidity (τ) at pH 5

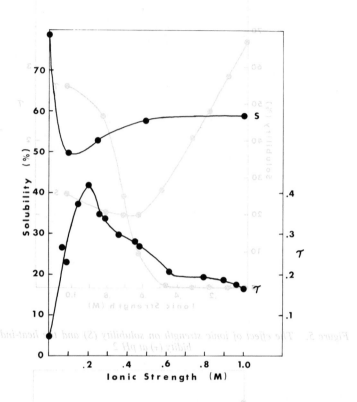

Figure 7.   The effect of ionic strength on solubility (S) and the heat-induced turbidity (τ) at pH 7

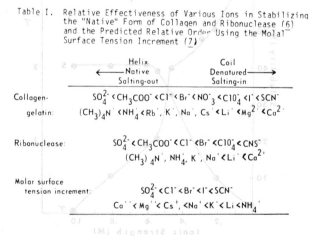

Table I.   Relative Effectiveness of Various Ions in Stabilizing the "Native" Form of Collagen and Ribonuclease (6) and the Predicted Relative Order Using the Molal Surface Tension Increment (7)

	Helix	Coil
	← Native	Denatured →
	Salting-out	Salting-in
Collagen-gelatin:	$SO_4^{2-} < CH_3COO^- < Cl^- < Br^- < NO_3^- < ClO_4^- < I^- < SCN^-$	
	$(CH_3)_4N^+ < NH_4^+ < Rb^+, K^+, Na^+, Cs^+ < Li^+ < Mg^{2+} < Ca^{2+}$	
Ribonuclease:	$SO_4^{2-} < CH_3COO^- < Cl^- < Br^- < ClO_4^- < CNS^-$	
	$(CH_3)_4N^+, NH_4^+, K^+, Na^+ < Li^+ < Ca^{2+}$	
Molar surface tension increment:	$SO_4^{2-} < Cl^- < Br^- < I^- < SCN^-$	
	$Ca^{++} < Mg^{++} < Cs^+, < Na^+ < K^+ < Li^+ < NH_4^+$	

enzyme activity, the stability of native globular and fibrillar structures, and the rates of conformation changes of proteins, polypeptides, and nucleic acids (6). These effects transcend in generality any details of macromolecule composition or conformation. Salts that are effective in increasing protein solubility (salting in) are also effective in destablizing native globular and fibrous structures and in increasing the rate of denaturation of native structures. Conversely, salts that decrease solubility (salting out) have the reverse effect on the stability and the rate of denaturation of native proteins. Because of this generality and the essentially independent effect of each ion, it is possible to arrange ions according to their effectiveness in salting out of proteins. This lyotropic or Hofmeister series (Table I) of ions was first compiled by Hofmeister in his studies on the salting out of euglobulins (7). Recent interpretations of these salt effects in terms of the basic hydrophobic and electrostatic forces between marcomolecules have given new insights into the behavior of proteins in solution.

Hermansson (4) has studied the effect of NaI, NaCl, $NaC_2H_3O_2$, Na_2SO_4 and $CaCl_2$ on the solubility of an isolate prepared under mild conditions. There has been no corresponding work reported for denatured soy isolates. Thus, what follows is an evaluation of the effects of neutral salts on the solubility of native and denatured soy isolates.

The native protein isolate (NPI) was prepared as follows. One part of commercial defatted soy flakes was extracted with ten parts of dilute NaOH (pH 9–10). After removal of spent flakes, the total extract was adjusted to pH 4.5 with dilute HCl to precipitate the curd. The curd was washed three times with H_2O, resuspended at pH 7.0, and freeze dried. Intrinsic viscosity, optical rotation, and differential scanning calorimetry (DSC) measurements show the soluble fraction of curd to be native.

The denatured protein isolate (DPI) was prepared as follows. Acid precipitated curd is washed, resuspended at pH 7.0, heated at temperatures above 90°C, and spray dried. DSC measurements indicate this isolate to be totally denatured. Both the NPI and DPI have approximately 1.5 to 2 percent residual ash, mainly NaCl.

Solubility measurements were those of Shen (1) except the centrifugation time is increased from 20 to 40 min. Under these conditions, particles with $S_{20,W} > 70$ were removed from the solutions.

The solubility data for NPI and DPI are plotted as a function of added salt concentration in Figures 8 and 9, respectively. Even though the NPI is native and the DPI is totally denatured, the solubility profiles of the two isolates are qualitatively very similar for all the salts studied. In solutions of the salts with monovalent anions, the solubility

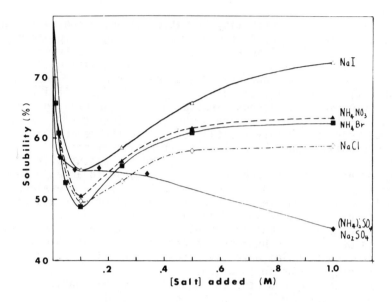

Figure 8.   The effect of various salts on the solubility of a NPI

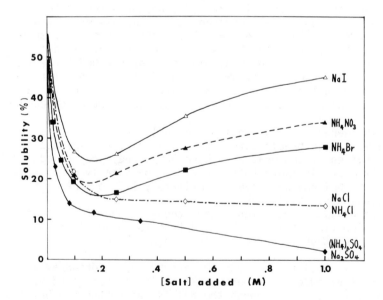

Figure 9.   The effect of various salts on the solubility of a DPI

profiles of both NPI and DPI exhibit a steep salting out at low
salt concentrations until minimum solubilities are reached at
0.1M and 0.2M salt for NPI and DPI, respectively. After that
the solubility increases gradually as the salt concentration is
increased to 1M. At this concentration, the net effect is
still salting out with the highest solubilities still well
below those in water. The magnitude of the initial salting out
and subsequent salting in is dependent upon the nature of the
salt. In general, the lyotropic order is observed with $Cl^- >$
$Br^- > NO_3^- > I^-$ in salting out efficacy.

In the case of the $SO_4^=$ anion, there is the initial
steep salting out. But there are plateaus between 0.1 and 0.3M
salt concentration instead of the minimum solubilities found in
the cases of the monovalent anions. After the plateau, the
solubilities decrease gradually with increasing salt concentration.

Quantitatively, these two isolates differ in their initial
solubilities in water and in the magnitude of the salt
effects. For a given salt, DPI is on the average salted out
twice as much as NPI. Our data agree qualitatively with those
of Hermansson (4) except for the lack of a minimum in the
Na_2SO_4 curve.

Effect of Salt on Other Proteins and Model Polypeptides

For comparison with these data, the types of salt effects
that have been observed for model protein and polypeptide
systems that achieve true thermodynamic equilibrium in solution
can be summarized into three classes:

Class I: Protein solubility is first increased (salting
in) and then decreased (salting out) as salt concentration
is increased. Thus, there is a maximum peak in the
solubility profile. Examples of proteins that show this
behavior are carboxylhemoglobin and fibrinogen (7).

Class II: Protein solubility is either monotonically
increased or decreased by increasing salt concentration.
The model polypeptide acetyltetraglycine ethyl ester
(ATGEE) behaves in this manner. Citrates, sulfates,
phosphates, acetates, and chlorides salt out ATGEE; whereas
phenol, perchlorates, tosylates, trichloroacetates,
thiocyanates, iodides, and bromides salt in (8). The heat
denaturation of ribonuclease also fits this class.

Class III: Protein solubility is decreased (salting out)
and is then increased (salting in) by increasing salt
concentration. Except for the soy isolates, there are no
other proteins to my knowledge that exhibit this type of
solubility behavior. However, in the related polymeriza-

tion and denaturation reactions, there are examples of this
type of behavior. Unpolymerized G-actin is caused to poly-
merize to F-actin by 0.1M NaI, but F-actin is depolymerized
by 0.5M NaI (9). The unfolding of the helical structure of
DNA also fits this class (6).

The Theory of Melander and Horvath (7, 10)

Recently, Melander and Horvath (7, 10) have proposed a
single theory to account for the effects of neutral salts on
the electrostatic and the hydrophobic interactions in the
salting out and the chromatography of proteins. In simplified
terms, the theory accounts for the solubility of proteins in
terms of two contributions, electrostatic and hydrophobic in
nature.

$$\ln (W/Wo) = (\Delta F_{electrostatic})/RT + (\Delta F_{cavity})/RT + Const$$

where W is the weight of protein soluble in 1 liter salt solution of given concentration and Wo is the weight of protein
soluble in 1 liter of H_2O.

The electrostatic term, $\Delta F_{electrostatic}$, is the change in
electrostatic free energy when the protein goes from the
crystalline state to the solution state. Melander and Horvath
have expressed this term by combining the proper terms from the
Debye-Huckel and Kirkwood theories (7). This term is always
positive, and is responsible for salting in.

The hydrophobic term, F_{cavity}, is the free energy
required to create a cavity in the bulk solvent to house the
hydrophobic groups that are exposed when a protein goes from a
crystalline state to a solution state.

$$\Delta F_{cavity} = -\Omega \sigma m \approx -\Delta A \sigma m$$

where Ω is a term proportional to the increase in hydrophobic
surface area (ΔA) as a protein goes into solution, σ is the
molal surface tension increment and m is the molality. The
molal surface tension increment, σ, is defined by the relationship: $\gamma = \gamma_0 + \sigma m$ where γ is the surface tension of the
salt solution and γ_0 is the surface tension of water.

For inorganic neutral salts, σ is positive. Thus, for
these salts, ΔF_{cavity} is always negative if hydrophobic
surface area is exposed upon solubilization (ΔA is positive).
This term accounts for the salting out of proteins.

The combination of the hydrophobic salting out and the
electrostatic salting in terms explains very nicely the Class I
type behavior. The data for carboxyhemoglobin and fibrinogen
are well fitted by the theory (Figure 10). Also the order of
decreasing molal surface tension increment generally follows
the lyotropic order (7).

The theory fails to explain the behavior of soy proteins
(Class III) because the salting in electrostatic term is more
dominant at low salt concentrations. As salt concentrations
increase, the cavity term becomes more dominant. The overall
result is salting in followed by salting out.

The failure of the theory indicates that soy protein
behavior is more complex than Melander and Horvath's (7, 10)
model. In deriving their simple theory, Melander and Horvath
assumed that the changes in hydrophobic surface area, in the
dipole moment, and in the net charge of the protein upon
solubilization are invariant with respect to salt species or
salt concentration. In other words, they assumed that the
soluble proteins have the same thermodynamic state regardless
of salt species or salt concentration. The results for soy
proteins can be treated in the framework of the Melander and
Horvath's theory if it is expanded to allow the exposed surface
area, the dipole moment, and the net protein charge to be
functions of the salt species and the salt concentration.
Thus, there are three different ways to account for this soy
protein behavior as follows:

1) Exposed surface is decreased by increasing salt concentration. The hydrophobic term is proportional to ΔA
   and m. If ΔA decreases more rapidly than the increase
   in m, the salting out term and, thereby, the degree of
   salting out, will decrease with increasing salt concentration. Since associated proteins (dimers, trimers,
   etc.) will have less exposed hydrophobic surface area
   than monomers, increasing amounts of associated
   proteins with increasing salt concentration will result
   in decreasing ΔA values. Soy proteins are known to
   associate with increasing salt concentration. At pH
   7.6, 11S and 7S proteins dissociate into subunits at
   ionic strength below 0.1M and 0.5M, respectively (11,
   12). This is, therefore, a plausible explanation of
   the observed solubility behavior. The sharp minimum at
   0.1M salt concentration in the case of the NPI might
   correspond to the cooperative association reactions
   involving 11S subunits. Even though there has been no
   experimental demonstration, it would not be unreason-
   able to postulate that random association reactions
   also occur for denatured soy proteins. The broader
   minimum between 0.1 and 0.2M salt concentration might
   be indicative of less cooperative association reactions.

2) Dipole moment increases with increasing salt concentration. The dipole moment of a protein is a function of
   the total net charge of the protein, the distribution
   of the charges on the protein, and the protein conformation. As salt concentration (ionic strength) is

increased, the solvent becomes increasingly better for
solvating charged species. Thus, it is plausible that
the protein coil might expand. This would result in
greater separation of charged sites on the protein and
a larger dipole moment. The increasing dipole moment
will make the electrostatic salting in term more
dominant as salt concentration increases. It is
difficult to determine whether this is an important
contribution in the cases studied here.

3) Ion binding causes net charge to go through point of
    zero net charge. If the net charge on the protein is
    reduced, the dipole moment should also be reduced. The
    dipole moment should be minimum at zero net charge. It
    is possible that there is a specific binding of counter
    ions that causes the protein to go through a point of
    zero net charge as salt concentration is increased. If
    this occurs, the electrostatic salting in term will
    decrease as salt concentration is increased and reach a
    minimum at a point where the net charge is zero. This
    will result in a minimum in the solubility profile.
    The effect of $CaCl_2$ on the solubility of soy proteins
    falls into this category (4). However, it is not
    expected that Na^+ or NH_4^+ will be appreciably
    bound by soy proteins. Thus, this effect should not be
    of great importance here.

In the context of this expanded theory, it is possible to
compare the exposed hydrophobic area of the native soy proteins
with that of the denatured proteins. It is evident from the
fairly large salting out effect of NPI at low salt concentrations that native soy proteins have appreciable amounts of
exposed hydrophobic surface area. If it is assumed that the
electrostatic term is negligible at low salt concentrations,
the initial slopes of the curves in Figure 11 are proportional
to the exposed hydrophobic surface area. Using the slopes it
is estimated that the exposed hydrophobic surface area of
native soy proteins is 50 percent of that of the denatured
proteins. This result is in agreement with the conclusion of
Takagi et al., (5), who used the binding of a fluorescent
hydrophobic dye to probe the hydrophobic surface of native and
heat denatured soy proteins. This is a rather gratifying
agreement considering the uncertainties involved in the
interpretations.

Viscosity

As was the case for solubility, apparent viscosity of
concentrated soy protein slurries is a non-equilibrium property
that is highly sensitive to past history of the slurry and to
the techniques of its measurement.

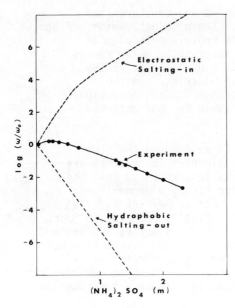

Archives of Biochemistry and Biophysics

Figure 10. Illustration of the fit of Melander and Horvath's theory to the experimental data for the solubility of carboxyhemoglobin (7)

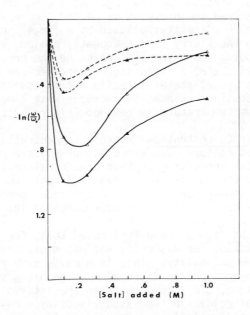

Figure 11. The solubility data for NPI (– – –) and DPI (——) in NaI ($\triangle$) and NH_4NO_3 ($\blacktriangle$) plotted as ln W/Wo

The intrinsic viscosity of native and denatured soy proteins have been measured (13). These values, which reflect the hydrodynamic properties of the protein molecules at infinite dilution, are of little interest to us as far as functionality is concerned. What is of interest is the apparent viscosity of concentrated slurries. In these slurries, the intermolecular protein–protein interactions dominate and are primarily responsible for the observed viscosity behavior.

Effect of Sample Preparation   It is well known that the viscosities of these concentrated slurries (8 percent) are shear rate dependent (14). What is less known is that the viscosity of denatured proteins are highly dependent upon dispersion conditions (Table II). For 12 percent slurries of Supro 620, viscosities decrease drastically from 10^4 to 33 cp as total shear is increased (15). If the slurries are allowed to stand, the viscosities increase slowly and revert to a much higher value (Table III). Thus, the highly sheared slurries are not at equilibrium. But, since the approach to equilibrium is slow, it is possible to use shear to produce a functionally desirable viscosity.

Progel Formation   Just as the viscosity can be lowered by shear, it can be increased by heating. As a concentrated slurry of undenatured soy proteins is heated, there is an initial decrease in viscosity followed by a large increase in viscosity. When the slurry is subsequently cooled, a rigid gel forms. Catsimpoolas and Meyer (16) describe the process in terms of a progel state, a fluid state characterized by high viscosity. This progel state is really not an equilibrium state. The viscosity continues to increase with time even after equilibrium temperature is reached (Figure 12).

Effect of Protein Concentration   Protein concentration is another important factor. The viscosity of soy proteins rises exponentially with increasing protein concentration (Figure 13). Further, different isolates have different exponential behavior. Such strong concentration dependence, progel and gel formation are evidences of strong intermolecular interactions.

Relation to Solubility   D. Chou (15) noted that, for a random sampling of isolates, the viscosity was not necessarily correlated to the solubility. This is because many factors (conformation, hydration, exposure of hydrophobic groups, charge distribution, etc.) contribute to the intermolecular interactions that result in increased viscosity. However, within a series of similarly processed isolates or with a given isolate, we should expect the viscosity to be inversely

Table II. The Effects of Dispersion Methods on the
Viscosity of 12 Percent Supro 620 Slurries
(15). Method A: polytron (Brinkmann) at
setting of 3.5 for 45 sec.; Method B:
blend with Osterizer blender set at "blend"
for 45 sec.; Method C: ultrasonicate with a
Braunsonic 1510 at 150 watts.

DISPERSION METHOD	BROOKFIELD APPARENT VISCOSITY (3 RPM)
A	1.0×10^4 cp
B	2.6×10^3 cp
C (90 sec.)	6.5×10^2 cp
C (7.5 min.)	33

Table II. Society of Rheology

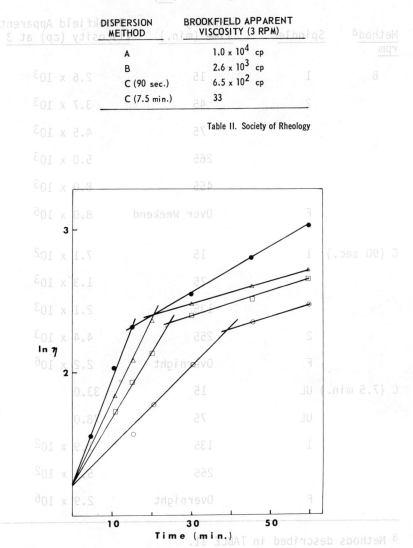

Cereal Chemistry

*Figure 12.   The viscosity of progel as a function of heating time at 65°C (○),
70°C (□), 75°C (△), and 80°C (●) (16)*

Table III. The Effect of Aging on the Apparent Viscosity of
            12 Percent Supro 620 (15)

Method[a] rpm	Spindle	Standing Time (min.)	Brookfield Apparent Viscosity (cp) at 3
B	1	15	2.6×10^3
	2	45	3.7×10^3
		75	4.5×10^3
		265	5.0×10^3
		455	8.0×10^3
	F	Over Weekend	8.0×10^6
C (90 sec.)	1	15	7.1×10^2
		75	1.3×10^3
		135	2.1×10^3
	2	255	4.4×10^3
	F	Overnight	2.2×10^6
C (7.5 min.)	UL	15	33.0
	UL	75	78.0
	1	135	1.9×10^2
		255	5.1×10^2
	F	Overnight	2.9×10^6

[a] Methods described in TABLE II.

Society of Rheology

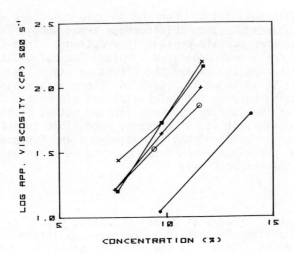

Society of Rheology

Figure 13.   The effect of protein concentration on viscosity (15). Sample solubilities (percent) are expressed in parenthesis.   A, (38) (×); B, (34) (+); C, (44) (○); D, (71) (■); laboratory-prepared sample (●).

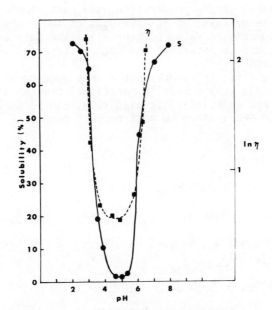

Figure 14.   The relation of viscosity (η) and solubility (S)

108 PROTEIN FUNCTIONALITY IN FOODS

correlated with solubility. For the soluble molecules with their larger exposed hydrophobic surface areas should experience greater protein-protein interactions. This is demonstrated for a heat denatured sample. The solubility is changed by altering the pH and the effect on viscosity is examined. As it is shown in Figure 14, there is a reasonably good correlation between the solubility and viscosity.

Because of the complex, non-equilibrium nature of the viscosity behavior, it is, at present, impossible to interpret this data in terms of the hydrodynamics and fundamental forces of soy proteins. Much more needs to be known about the dynamics of the interactions.

Conclusion

As functional properties of soy proteins, viscosity and solubility are alike in that they are non-equilibrium properties of the system. In the case of solubility, there is at least evidence of steady state equilibrium which allows for the possibility of some qualitative thermodynamic interpretation. In the case of viscosity, steady state equilibrium is not reached. Thus, thermodynamic interpretation is impossible. Molecular dynamics data are needed.

Because solubility and viscosity are non-equilibrium properties, the measured values are dependent upon measurement conditions, sample preparation, and past sample history. Because of this, it is difficult or sometimes even impossible to compare data, to generalize the data and make predictions, and to interpret the data in terms of the basic molecular forces.

Soy protein is, therefore, not a very good model protein. However, it is a very versatile functional protein in that its solubility, and especially its viscosity, can be tailored by processing or fabrication to meet product needs.

Literature Cited

1. Shen, J. S. Cereal Chem., 1976, 53, 902.

2. Wolf, W. J.; Tamura, T. Cereal Chem., 1969, 46, 331.

3. Catsimpoolas, N., Funk, S. K., Meyer, E. W. Cereal Chem., 1970, 47, 331.

4. Hermansson, A-M. J. Text. Studies, 1978, 9, 33.

5. Takagi, S., Nagaoki, O., Motonari, A.; Yasumatsu, K. Nippon Shokuhin Kogyo Gakkaishi 1979, 26(3), 139.

6. Von Hippel, P. H.; Schleich, T. in "Structure and Stability of Biological Macromolecules," Vol. 2, Eds. Timasheff, S. N.; Fasman, G. D. Dekker, 1969, p 417–574.

7. Melander, W.; Horvath, C. Arch. Biochem. Biophys., 1977, 183, 200.

8. Robinson, D. R.; Jencks, W. P. J. Am. Chem. Soc., 1965, 87(11) 2470.

9. Nagy, B.; Jencks, W. P. J. Amer. Chem. Soc., 1965, 87(11), 2480.

10. Melander, W.; Horvath, C. J. Solid–Phase Biochem. 1977, 2(2), 141.

11. Wolf, W. J.; Briggs, D. R. Arch. Biochem. Biophys., 1958, 76, 377.

12. Koshiyama, I. Agr. Biol. Chem. (Tokyo) 1968, 32, 879.

13. Shen, J. S. J. Agric. Food Chem. 1976, 24(4), 784.

14. Circle, S. J.; Meyer, E. W.; Whitney, R. W. Cereal Chem. 1964, 41, 157.

15. Chou, D.; Richert, S. H. Golden Jubilee Meeting, Society of Rheology, Nov. 1979, Boston, MA. paper 32-3.

16. Catsimpoolas, N.; Meyer, E. W. Cereal Chem. 1970, 47, 559.

RECEIVED September 5, 1980.

6

Adhesion and Cohesion

J. S. WALL and F. R. HUEBNER

Northern Regional Research Center, Science and Education Administration,
Agricultural Research, U.S. Department of Agriculture, Peoria, IL 61604

Cohesion and adhesion are essential functional properties
of certain of the constituents of food mixtures if we wish to
convert them to sight-appealing shaped products with acceptable
textures. Parker and Taylor (1) define adhesion as the use of
one material to bond two other materials together and cohesion
as the joining together of the same material. The development
of textured foods based on vegetable proteins, multi-component
breakfast foods, and some prepared meat specialty items has
required at least one ingredient, usually protein, to serve as
binder to hold components together. The binding agents may
function before or after cooking the ingredient mix; cooking
establishes additional adhesive and cohesive interactions
among protein, lipid, and carbohydrate components of foods.
      In this paper, we will explore the measurement of and the
basis for the cohesive and elastic properties of a commonly
used component of foods that excels in these characteristics,
wheat gluten. Gluten constitutes from 10 to 16% of wheat
flour, from which it may be separated by Martin, batter, or
Raisio processes (2, 3). The separated wheat gluten is 70 to
80% protein, of which 85% is insoluble in saline solution. We
shall also seek to correlate some of the basic concepts developed
in studies of gluten to other protein systems, such as those
of soybean protein isolates and concentrates.
      A good example of the contribution of protein to adhesion
and cohesion of a multi-component system is wheat flour dough.
Khoo et al. (4) have observed with the scanning electron
microscope that dough consists of starch granules held together
by a matrix of hydrated gluten protein, which is stretched
into coherent films (Figure 1). These films are not artifacts
of microscopy since isolated gluten is an excellent film
former. Whole gluten purified by solubilization in dilute
acids can be dispersed in lactic acid, which acts as a humectant
and plasticizer, and cast as a film as shown by Wall and
Beckwith (5). Such films would serve as edible coatings.
Film-making properties are characteristic of many polymers,

including proteins, and are a good measure of cohesive strength.
Gluten proteins have also been tested as industrial adhesives
for bonding paper or wood (5). Just as gluten proteins associate
with starch, so does it bond to other polar materials, such as
cellulose. Many of the concepts developed in industrial
adhesives apply to food applications.

Experimental Procedures

Many instruments are available to provide fundamental
information on the adhesive and cohesive properties of food
products of different forms and at different stages of preparation. Voisey and de Man (6) classify such instruments into
two main categories: (a) linear motion instruments which
generally measure extensibility and tensile strength or compression and flexing strength, and (b) rotary motion devices which
measure resistance to flow or mixing of viscous solutions or
plastic masses. The linear motion instruments usually rupture
the structures and so are best used to measure a specific
stage of product development. In contrast, the rotary instruments
often may be used to examine transitory effects on physical
properties induced by temperature variation, chemical additives,
or stresses caused by mixing.

The most widely used linear motion analyzers are the
Instron Universal Testing machines which range from large
floor models to table instruments suitable for most food tests
(7, 8). Force is applied by drive screws to a crosshead which
is moved downward at a specified rate. An upper force measurement
cell measures extending forces on materials clamped between it
and the moving crosshead (Figure 2). Wall and Beckwith (5)
used this system to measure adhesion by wheat gluten in model
systems. Data on force and distance traversed by the crossarm
are recorded. Frazier et al. (9) used the Instron to compress
dough or gluten balls at a given load and measure time required
for the balls to relax to a lower load at a constant deformation.
Various attachments to the Instron permit measurement of
lateral stresses such as flexing. Rasper (10) has modified
the Instron to permit dough extension measurements in a
temperature-controlled chamber containing fluid with density
equal to that of dough. Instruments specifically designed for
dough extensibility and tensile strength measurements are the
Brabender Extensograph or Simon Extensometer (10). In the
extensograph, a rod-shaped mass of dough is clamped horizontally
and a hook extends it vertically at a constant rate. Resistance
and extension are recorded until the dough cylinder breaks.
The area under the curve and the ratio of resistance to extensibility provide useful information as to the elasticity of the
dough.

Hydrated food products exhibiting adhesive or cohesive
properties are generally highly viscous or plastic and, therefore,

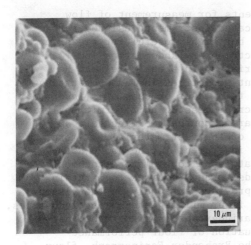

Figure 1.   SEM of wheat flour dough after mixing (4)

Figure 2.   Tensile strength measurement with the Instron Universal Testing machine

require special rotary instruments for measurement of flow
properties. A highly sophisticated instrument designed for
elastomer and plastic research but highly useful for research
on foods is the Mechanical Spectrometer (Rheometrics Inc.,
Union, NJ). A considerable variation in shear rates can be
applied in a number of different geometries including rotating
cone and plate between which a thin layer of sample is inserted.
The mechanical assembly and microprocessing unit permit analysis
of a broad spectrum of molecular responses yielding data on
viscous and elastic properties of the test material. A more
generally available laboratory instrument is the Haake Roto-
visco viscometer which also features a cone-plate attachment.
This modified instrument records torque required to maintain
constant rate of rotation and provides data only on the viscous
component of flow (8). In the flour milling and baking industries,
recording mixers, designated farinographs or mixographs, are
widely used for empirical evaluation of flour performance
during dough processing. In the Brabender Farinograph, flour
and water are mixed by revolving blades. The motor shaft is
connected to a dynamometer which measures resistance to the
mixing (11, 12). The chart records the dynamic process of
dough making and breakdown with time. Water content (absorption)
is adjusted to give similar resistances at maximum dough strength.
The mixograph is similar in concept to the farinograph but uses
pins instead of blades and imposes a greater stress on the
dough (11).
    While these methods can provide useful information for
determining the functional performance of ingredients used in
food, the criteria for quality must be established on the final
product; thus, dough must be baked into bread and meat analogues
cooked and these foods subjected to organoleptic evaluation for
texture.

Molecular Interactions

    Extensive studies in protein chemistry (13) and synthetic
polymers (14) have established compositional factors responsible
for molecular associations involved in adhesion and cohesion
phenomena of proteins and other polymers. Figure 3 summarizes
types of functional groups in proteins participating in associ-
ative interactions and agents that disrupt the bonds they form.
    Electrostatic charges due to ionized acidic or basic amino
acids influence protein solubility. At extremes of pH, many
poorly soluble proteins are dissolved and their molecular
structures unfolded due to surplus of similar repelling charges.
Gluten proteins have few charged groups and so are poorly
soluble in neutral solution (15). Dispersions of other proteins
must be adjusted to their isoelectric point or have salt added
to optimize cohesion and adhesion.

Polar groups contribute greatly to adhesion of proteins to carbohydrates and to their cohesion. In denatured or unfolded proteins, such as animal glues, the peptide amide groups play an important role in adhesion; but in the undenatured collagen, most peptide groups are associated in helical conformations. In undenatured proteins, side chain amide groups from the amino acids glutamine and asparagine and hydroxyl groups of serine and threonine interact through hydrogen bonds. Gluten proteins contain over 33% glutamine and asparagine in their amino acid composition (15).

Investigators of the chemistry of adhesion refer to nonpolar interactions involving long chain aliphatic or aromatic groups in terms of Van der Waal or London forces (1). Protein chemists generally use the term "hydrophobic bonding" to describe these interactions, because in aqueous systems nonpolar residues in proteins tend to retreat and associate in the molecule's interior or with other like groups on adjacent molecules. Membrane proteins especially have exposed hydrophobic groups which contribute to association with lipids and integrity of the membrane structure.

Disulfide bonds in the amino acid cystine are important to the properties of many proteins by maintaining covalent intramolecular bonds and crosslinks between protein chains (16).

The opposing effects of hydrogen bonding and negatively charged amino acid side chains on molecular aggregation of proteins have been demonstrated by experiments with model systems (17). Synthetic polypeptides were prepared containing both polar hydrogen bond–forming glutamine residues and glutamic acid groups. As shown in Figure 4, 2M and 8M urea helped solubilize the polymers in aqueous solutions at low pH by dissociating hydrogen bonds between amide groups. But as the fraction of glutamine residues was increased in the polypeptides, it was necessary to raise the pH to induce more negative electrostatic charges on the glutamic acid residues in order to dissociate the polypeptides held together by hydrogen bonds.

The participation of hydrophobic groups of native wheat gluten proteins in intermolecular associations was demonstrated by Chung and Pomeranz (18) through use of hydrophobic gels as illustrated in Figure 5. These workers introduced a solution of gluten proteins in 0.01M acetic acid into a column of Phenyl-Sepharose-4B. Little protein was eluted by washing the column with dilute acetic acid, but about 40% of the protein was eluted by a solution containing 1% of the detergent sodium sodium dodecyl sulfate. Only a small amount of additional protein was eluted by other solvents including 0.005M glycine-NaOH in 50% propylene glycol. Thus, although hydrophobic bonds are individually weak, their combined strength in proteins, which can provide many such interactions, can be considerable.

The role of noncovalent bonds in determining protein structure and aggregation has been confirmed by x-ray analysis

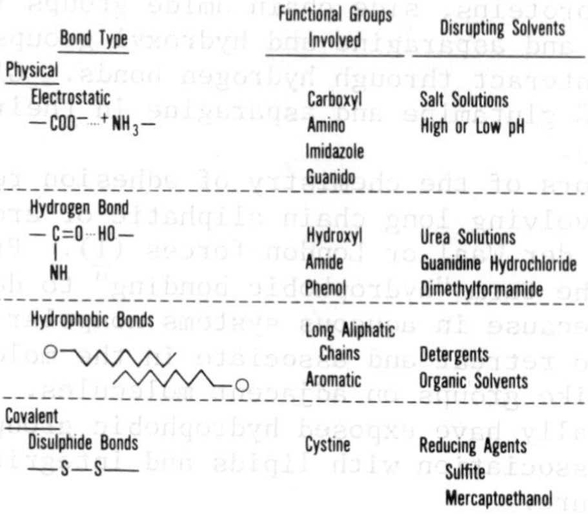

116   PROTEIN FUNCTIONALITY IN FOODS

Bond Type	Functional Groups Involved	Disrupting Solvents
Physical		
Electrostatic	Carboxyl	Salt Solutions
— COO⁻ ⁻⁺NH₃ —	Amino	High or Low pH
	Imidazole	
	Guanido	
Hydrogen Bond		
— C=O··HO —	Hydroxyl	Urea Solutions
|	Amide	Guanidine Hydrochloride
NH	Phenol	Dimethylformamide
Hydrophobic Bonds		
◯〜〜〜	Long Aliphatic Chains	Detergents
〜〜〜◯	Aromatic	Organic Solvents
Covalent		
Disulphide Bonds	Cystine	Reducing Agents
— S — S —		Sulfite
		Mercaptoethanol

Figure 3.   *Types of bonds between protein chains*

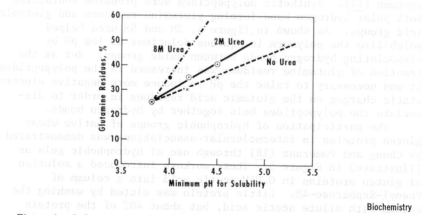

Biochemistry

Figure 4.   *Influence of urea and pH on the solubilities of synthetic polypeptide copolymers of glutamine and glutamic acid (17)*

of protein crystals. At 2Å resolution, the sites of the individual amino acids and the proximities of their functional groups
can be deduced, and the nature of the bonding forces involved
in maintaining the folding and association of polypeptide
chains can be established. Studies on the structure of the
molecules of the enzyme trypsin, the soybean trypsin inhibitor,
and the complex between the two proteins indicate that multiple
types of noncovalent linkages involving hydrogen bonds, hydrophobic
bonds, and electrostatic attractions participate in the molecular association (19).

Molecular Size and Shape

Not only do the kinds and amounts of functional groups on
proteins govern the extent of protein interactions, but their
location on the molecule and their accessability to groups in
other molecules are important also. These factors depend on
the size and shape of the protein molecules. Polymer and
adhesive chemists concur that highly cohesive films and other
structures are attained by high-molecular-weight molecules that
allow extensive molecular interactions and by those with numerous
intermolecular covalent crosslinks (1, 14). In contrast,
adhesion depends more on accessibility of functional groups of
the adhesive to the adhering materials.
     As shown in Figure 6, proteins of the gluten complex can
be separated by solubility differences into saline-soluble
albumins and globulins, 70% ethanol-soluble gliadins, acetic
acid-soluble glutenin, and an insoluble protein residue.
Albumins and globulins are highly folded compact molecules.
Much of their backbone chain is involved in helical hydrogen-
bonded associations. Folding is maintained by both internal
hydrogen and hydrophobic bonds as well as by disulfide bonds.
The gliadins, with few charged groups, associate to yield a
syrupy mass on hydration. But the large asymmetric glutenin
molecules form a tough, rubbery, cohesive mass when hydrated.
The large size of soluble glutenin molecules is due to limited
disulfide bonds between polypeptide chains. The insolubility
of residue protein is attributable to extensive intermolecular
disulfide crosslinks.
     Evidence for variation in molecular size of gluten proteins
was obtained when protein extracts of flour were chromatographed
on agarose gel filtration columns (Sephadex Cl-4B) in tris
buffer containing sodium dodecyl sulfate as shown in Figure 7
(20). Most albumins (Alb) and globulins (Glob) as well as
gliadins elute late, indicating they have molecular weights
below 40,000. In contrast, glutenin shows a broad spectrum of
species with some components having molecular weight over
1 million. Use of the hydrophobic bond-breaking solvent sodium
dodecyl sulfate disrupted aggregation of the proteins. The

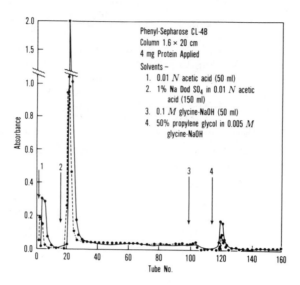

Cereal Science Today

Figure 5.   Elution of acid-soluble wheat proteins from hydrophobic gel Phenyl-Sepharose-4B by different solvents. NaDodSO$_4$ = SDS (18).

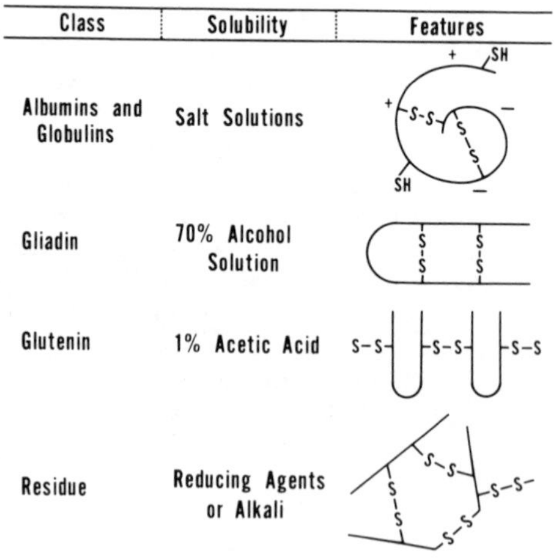

Class	Solubility	Features
Albumins and Globulins	Salt Solutions	
Gliadin	70% Alcohol Solution	
Glutenin	1% Acetic Acid	
Residue	Reducing Agents or Alkali	

Figure 6.   Types of proteins in wheat flour as separated by solubility

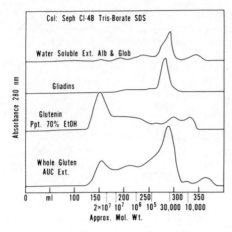

Journal of Agricultural and Food Chemistry

Figure 7. Fractionation of wheat proteins on sepharose-CL4B columns by gel filtration chromatography in 0.125 tris-borate buffer, pH 8.9 and 0.1% SDS (20). AUC ext. refers to protein extracted from wheat flour with a solution containing 0.1M acetic acid, 3M urea, and 0.01M cetyltrimethyl-ammonium-bromide.

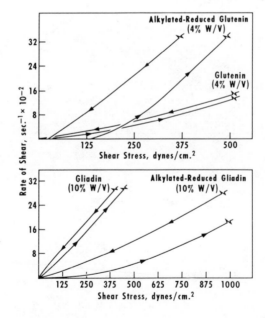

Cereal Science Today

Figure 8.   Shear–stress curves for native and alkylated–reduced gliadin and glutenin solutions at low shear stresses (5)

low-molecular-weight glutenin protein fraction was not separated
in other solvents and appears to consist of membrane proteins
that tend to aggregate.

The effect of molecular size and shape on protein cohesive
strength was indicated by measurements of tensile strength and
elongation of films cast from laboratory preparations of wheat
gluten, gliadin, and glutenin (5). The measurements were made
with a Scott Tester Model IP-2 at 20°C and 42% relative humidity.
Values for force applied to the film (tensile strength) and
elongation at time of film rupture are listed in Table I.
Glutenin, which consists of the larger, more asymmetric molecules,
forms films with greater tensile strength than gliadin films.
Gliadin films stretch further than those from glutenin due to
their having weaker molecular associations. Gluten, which is a
mixture of gliadin and glutenin, has intermediate film properties.

Marked changes in the viscosities of glutenin and gliadin
solutions, as measured by cone-plate viscometer, occur after
cleavage of their disulfide bonds by a reducing agent (Figure 8).
Native glutenin has high viscosity, which indicates not only
high molecular weight but also a highly asymmetric structure
(5). Furthermore the hysteresis or deviation of the viscosity
curves for increasing and decreasing shear stress versus rate
of shear provide evidence for non-Newtonian behavior of these
molecules due to molecular interactions. Glutenin viscosity

TABLE I

Tensile Strength and Percent Elongation

of Wheat Protein Films

Film	Tensile strength $\mathrm{lb/in^2 \times 10^{-3}}$	Elongation %
Glutenin	3.38	63
Gliadin	1.42	72
Whole gluten (laboratory preparation)	1.75	75

From Wall and Beckwith (5).

declines markedly when its intermolecular disulfide bonds are
cleaved to liberate its constituent polypeptide chains. The
viscosity of native gliadin in solution is much less than that
of glutenin. Reduction of the disulfides of gliadin results in
a significant increase in its viscosity due to unfolding of the
polypeptide molecule. Reduction of glutenin destroys its
cohesive nature when hydrated, but the reduced proteins are
very sticky and quite adhesive.

Dough Rheology

Because of the gluten proteins, hydrated flour can be
worked into an elastic-cohesive mass by mixing. The development of optimal dough properties with time during the mixing
process can be followed on a mixograph. The mixograph recordings
in Figure 9 show that initially the unoriented dough molecules
offer little resistance. As mixing proceeds, the asymmetric
glutenin molecules are oriented and associate to increase dough
strength. Disulfide-sulfhydryl interchanges to facilitate
rearrangement of the molecule and actual cleavage of the disulfide links may also take place during mixing (23). Finally,
resistance to mixing declines as polymer disruption continues.
The three curves show different mixing responses from flours of
wheats of different breadmaking quality as measured by Finney
et al. (21) and Finney and Shogren (22). The middle curve
shows resistance changes with time for dough from a flour with
suitable properties for breadmaking. It exhibits a moderate
dough development time and stability time. The upper curve is
that for dough from a wheat whose gluten is overly strong,
since it requires longer mixing to achieve maximum dough strength,
whereas the lower curve is of a weak flour dough with short
mixing time requirement but rapid breakdown of dough strength.
    The relationship between different flours varying in dough
strength and their composition of different protein fractions
was investigated by Orth and Bushuk (24) and by Huebner and
Wall (25). The former workers found a correlation between the
mixing time requirement of doughs and their tolerance to mixing
to their content of residue protein. As shown in Figure 10,
the latter workers analyzed for protein content a series of
flours derived from different Hard Red Winter wheat varieties
that vary in mixing time requirement (mixing strength). The
stronger flours contained not only more residue protein but
also more of the higher molecular weight glutenin fraction
(Glutenin I).
    Changes occurring in disulfide bonds of wheat gluten
during dough mixing are supported by two observations. Mecham
and Knapp (26) found that mixing in the absence of air increases
the content of sulfhydryl groups in the dough. Also, there is
an increase in extractable protein during mixing. This increase
in extractable protein is primarily high-molecular-weight

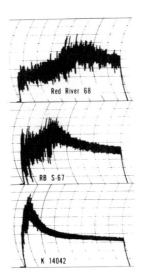

Figure 9. Mixograms of doughs from flours prepared from three varieties of wheats varying in dough performance (21, 22)

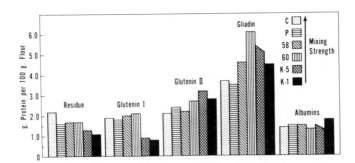

Cereal Chemistry

Figure 10. Yields of protein fractions isolated from hard red winter wheat flours differing in mixing strengths (25)

protein according to gel filtration studies conducted by Tsen
(27), whose data is given in Figure 11. Apparently, mixing
breaks down disulfide bonds in the highly crosslinked residue
protein, thereby decreasing resistance to mixing. Addition of
reducing agents such as cysteine which cleave disulfide bonds
weakens the dough (21), whereas addition of oxidizing agents
such as bromate tends to strengthen it by eliminating sulfhydryl
groups (26).

Figure 12 summarizes the contribution of various gluten
proteins to dough properties. The large asymmetric glutenin
molecules have considerable surfaces with numerous exposed
functional groups to permit strong association by noncovalent
forces. Fragments of highly crosslinked residue proteins
contribute lateral cohesion and resistance to laminar flow.
During mixing the residue proteins are probably degraded to
yield linear, more soluble molecules. The smaller gliadin
molecules are less tightly bound and facilitate fluidity and
expansion of the dough. These ideas are consistent with experimental findings by Hoseney et al. (28), who separated the
gluten proteins from flours of different baking quality and
substituted gliadin or glutenin protein from good baking wheat
flours for the same protein fraction in poor flours in recon-
stituted doughs. The gliadin fraction from good wheat flours
appeared to improve loaf volume, while the glutenin fraction
affected the mixing requirement and tolerance. All components
of dough must be present in proper amounts for good breadmaking
properties. If the protein is too cohesive and tough, the
dough will not rise properly because expansion of trapped
yeast-generated CO_2 bubbles will be minimal; but, if the protein
matrix is weak, the gas pockets will break and the dough will
collapse.

Uses of Wheat Gluten

The unique properties of wheat gluten proteins have resulted
in considerable use of isolated gluten preparations. Of the
20 million kg used in the United States, 69% is used in bakery
products, 12% in breakfast foods, 9% in pet foods, and 4% in
meat analogs (2). In baked goods, gluten may be used to
supplement weak flours to provide additional mixing strength
and tolerance. It may be used in specialty products, such as
high-fiber breads, where the added gluten provides better loaf
volume. A major use is in production of hamburger buns, where
the supplemented gluten increases the structural strength of
the hinge. When hydrated gluten is heated above 85°C, the
protein is denatured but retains its shape and its resiliency
(29). In bread and rolls, the gluten helps retain moisture in
the crumb and contributes to crumb strength.

Gluten is used in many other foods where its adhesive and
cohesive properties provide beneficial value (29). In breakfast

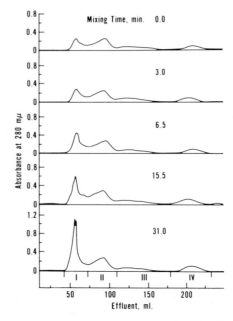

Cereal Chemistry

Figure 11.   Gel filtration of acetic acid extracts of flour and of dough mixed different times (27)

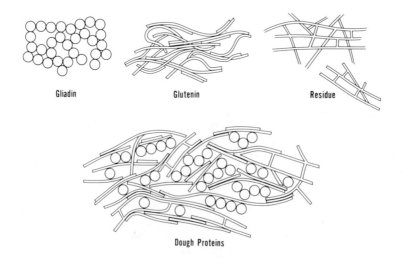

Figure 12.   Effect of wheat protein structures on molecular associations and visco-elastic properties

cereals, it is used to bond the formulation prior to cooking. The cohesive-elastic character of gluten is the basis for many vegetarian-simulated meat products where it holds ingredients and provides chewy texture. It must be cautioned that, because of its highly cohesive properties, native gluten may not be compatible with and serve as an adhesive for some ingredients. In the case of meat chunks, however, gluten appears to be one of the best binding agents among nonmeat proteins. As shown in Table II, Siegel and his associates (30) have tested the adhesive strength of several protein dispersions in salt and phosphate solution for binding meat particles. The bonded meat cubes were heated to 75°C before slicing, and tensile strength measurements were conducted on them. Only gluten and egg whites were better than the salt solution controls in cementing the meat pieces together.

Thermal and Alkaline Improvement of Adhesion in Globular Proteins

Most globular or albumin plant proteins exhibit little cohesive or adhesive properties in their native state. At higher pH, 11 or above, disulfide bonds are cleaved, protein unfolding occurs, and functional groups previously associated within the molecule become available for external binding. The paper sizing and plywood industries use strongly alkaline dispersions of soybean proteins as industrial adhesives. But for food use, drastic alkaline treatment is not desired for it results in loss of cystine and formation of lysinoalanine (31). However, milder alkaline treatment is used to denature soybean proteins prior to spinning fibers (32). The protein cohesion resulting when the fibers are coagulated into acid and salt is necessary to maintain fiber integrity (Figure 13). In the illustrated product "Bacos," egg albumin is used as an adhesive to bond the fibers into a desired fabricated meat-like product (33).

Heat denaturation is the most widely used and most important means of altering the adhesive and cohesive properties of globular proteins. Extrusion cooking is growing in use as a means of simultaneously shaping, texturizing, and stabilizing protein-rich products. Jeunink and Cheftel (34) have studied the nature of the changes in the protein that follow extrusion cooking of field bean protein concentrate. As shown in Figure 14, the protein in the neutral concentrate is 80% soluble in neutral phosphate buffer; but after extrusion, only 20% is extracted by that solution. Extrusion denatured much of the protein; unfolding of the chain permitted new functional group associations which rendered it insoluble. In the presence of the dissociating solvent, sodium dodecyl sulfate (SDS), and the disulfide breaking agent, dithiothreitol (DTT), most of the protein is solubilized. The aggregation of the extruded protein evidently is maintained by hydrophobic bonds and intermolecular disulfide links produced during heating.

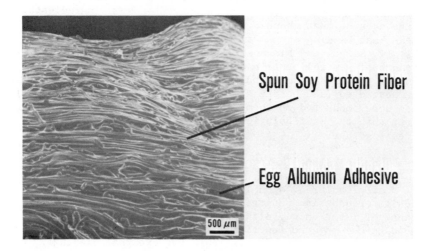

*Figure 13.   SEM of Bacos spun soybean protein fiber simulated meat product.
Egg albumin used as adhesive for fibers (33).*

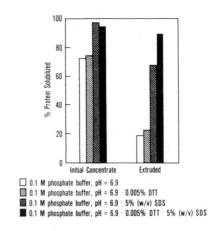

*Figure 14.   Protein solubility of initial
and extruded field bean protein concentrates in different extraction solutions (26)*

Cereal Chemistry

TABLE II

Meat Binding Abilities of Various Nonmeat

Proteins in the Presence of 8% Salt and 2% Phosphate

Protein	Binding ability, grams
Wheat gluten	175.4
Egg white	120.3
Control	107.0
Calcium reduced dried skim milk	74.5
Bovine blood plasma	71.9
Isolated soy protein	66.7
Sodium caseinate	0

Journal of Food Science

Cooking and texturizing vegetable proteins appears desirable if their properties for use in frankfurters or other protein-supplemented meat products are to be optimized. Uncooked soybean or cottonseed proteins do not maintain the texture of prepared frankfurters when added at 10% or 30% levels to the formulations according to Terrell et al. (35). Stress deformation of frankfurters was measured on the Universal Instron Testing machine with a L.E.E.-Kramer press. All-meat frankfurters showed higher values of stress-deformation prior to rupture than products with added soy flour, soy concentrate, liquid-cyclone-processed cottonseed flour, or isolated cottonseed protein. The textured soy flour and textured cottonseed protein performed better than the uncooked proteins; at the 30% level, the addition of textured soy proteins resulted in frankfurters having a higher deformation strength than the all-meat one.

Conclusions

Adhesion and cohesion are properties of many polymeric substances including proteins. The effectiveness of the proteins in bonding or shaping food ingredients is dependent on their composition and structure. Hydrophobic and hydrogen bonding functional groups on amino acids associate with like groups within the protein to influence conformation or between molecules to result in aggregation. Disulfide bonds between proteins result in larger molecules or insoluble complexes. High molecular weight and random coil structure of protein result in more associations and thereby enhance adhesive and cohesive properties. Although these characteristics are inherent in native gluten proteins, functional properties of other proteins may be improved by chemical or thermal processing. There is no universally good adhesive for food constituents. Proteins that are highly cohesive may not blend well with certain other ingredients. It is necessary to examine the available proteins for optimum properties and to select the most satisfactory ingredient combinations. A number of instruments are available for measurements of textural properties of food ingredients or products, but the final criteria for acceptable performance must be taste-panel evaluations.

The mention of firm names or trade products does not imply that they are endorsed by the U.S. Department of Agriculture over other firms or similar products not mentioned.

Literature Cited

1. Parker, R. S. R.; Taylor, P. "Adhesion and Adhesives;"
     Pergamon Press, London, 1966.

2. Sarkki, M. L.   J. Am. Oil Chem. Soc., 1979, 56, 443.
3. Wookey, N.   J. Am. Oil Chem. Soc., 1979, 56, 306.
4. Khoo, U.; Christianson, D. D.; Inglett, G. E.   Bakers Dig.,
    1975, 49(4), 24.
5. Wall, J. S.; Beckwith, A. C.   Cereal Sci. Today, 1969,
    14(1), 16.
6. Voisey, P. W.; de Man, J. M.   Applications of Instruments
    for Measuring Food Texture, in "Rheology and Texture in
    Food Quality," J. M. de Man, P. W. Voisey, V. F. Rasper,
    and D. W. Stanley, Eds., Avi Publishing Co., 1975, p. 142.
7. Bourne, M. C.; Mayer, J. C.; Hand, D. B.   Food Technol.,
    1966, 20, 170.
8. Voisey, P. W.   Instrumental Measurement of Food Texture,
    in "Rheology and Texture in Food Quality," J. M. de Man,
    P. W. Voisey, V. F. Rasper, and D. W. Stanley, Eds., Avi
    Publishing Co., 1976, p. 79.
9. Frazier, P. J.; Daniels, N. W. R.; Eggitt, P. W. R.
    Cereal Chem., 1975, 52 (3, part II), 106r.
10. Rasper, V. F.   Cereal Chem., 1975, 52 (3, part II), 24r.
11. Shuey, W. C.   Cereal Chem., 1975, 52 (3, part II), 42r.
12. Shuey, W. C.   "Farinograph Handbook." American Association
    of Cereal Chemists, St. Paul, Minnesota, 1972.
13. Anfinsen, C. B.; Scheraga, H. A.   Experimental and Theo-
    retical Aspects of Protein Folding; in "Adv. Protein
    Chemistry," 29, C. B. Anfinsen, J. T. Edsall, and F. M.
    Richards, Eds., Academic Press, New York, 1975, p. 205.
14. Van Krevelan, D. W.   "Properties of Polymers, Their
    Estimation, and Correlation with Chemical Structure;"
    Elsevier, Amsterdam, 1976.
15. Krull, L. H.; Wall, J. S.   Bakers Dig., 1969, 43(4), 30.
16. Wall, J. S.   J. Agric. Food Chem., 1971, 19, 619.
17. Krull, L. H.; Wall, J. S.   Biochemistry, 1966, 5, 1521.
18. Chung, K. H.; Pomeranz, Y.   Cereal Chem., 1979, 56, 196.
19. Sweet, R. M.; Wright, H. T.; Janin, J.; Clothia, C. H.;
    Blow, D. M.   Biochemistry, 1974, 13, 4212.
20. Huebner, F. R.; Wall, J. S.   J. Agric. Food Chem., 1980,
    28, 433.
21. Finney, K. F.; Tsen, C. C.; Shogren, M. D.   Cereal Chem.,
    1971, 48, 540.
22. Finney, K. F.; Shogren, M. D.   Bakers Dig., 1972, 46(2),
    32.
23. Axford, P. W. E.; Campbell, J. D.; Elton, G. A. H.   Chem.
    Ind. London, 1962, 407.
24. Orth, R. A.; Bushuk, W.   Cereal Chem., 1972, 49, 268.
25. Huebner, F. R.; Wall, J. S.   Cereal Chem., 1976, 53, 258.
26. Mecham, D. K.; Knapp, C.   Cereal Chem., 1966, 43, 226.
27. Tsen, C. C.   Cereal Chem., 1967, 44, 308.
28. Hoseney, R. C.; Finney, K. F.; Shogren, M. D.; Pomeranz,
    Y.   Cereal Chem., 1969, 46, 126.

RECEIVED September 5, 1980.

29. Kalin, F. J. Am. Oil Chem. Soc., 1979, 56, 477.
30. Siegel, D. G.; Church, K. E.; Schmidt, G. R. J. Food
    Sci., 1979, 44, 1276.
31. DeGroot, A. P.; Slump, R. J. Nutr., 1969, 98, 45.
32. Smith, A. K.; Circle, S. J. Protein Products as Food
    Ingredients in "Soybeans: Chemistry and Technology I;"
    A. K. Smith and S. J. Circle, eds. Avi Publishing Co.,
    Westport, Connecticut, 1972, p. 339.
33. Wolf, W. J. Scanning Electron Microscopy Studies of Soybean
    Proteins; "IIT Symposium on Scanning Electron Microscopy"
    1980 (in press).
34. Jeunink, J.; Cheftel, J. C. J. Food Sci., 1979, 44, 1322.
35. Terrell, R. N.; Brown, J. A.; Carpenter, Z. L.; Mattill,
    K. F.; Monagle, C. W. J. Food Sci., 1979, 44, 865.

Gelation and Coagulation

RONALD H. SCHMIDT

Food Science and Human Nutrition Department, University of Florida, Gainesville, FL 32611

The structural characteristics of a variety of food systems are complexly related to the physicochemical protein phenomena of aggregation, coagulation and/or gelation. These phenomena are physical manifestations of protein denaturation processes which are highly dependent upon the type and amount of protein, processing conditions, pH and ionic environment.

A complex interrelationship between association-dissociation, precipitation, coagulation and gelation reactions exists in protein systems (1,2). Protein association reactions generally refer to changes occurring at the molecular or subunit level while aggregation reactions generally involve the formation of higher molecular weight complexes from association reactions. The terms gelation and coagulation are not as clearly defined. Gelation may be theoretically defined as a protein aggregation phenomenon in which polymer-polymer and polymer-solvent interactions and in which attractive and repulsive forces are so balanced that a well ordered tertiary network or matrix is formed. This matrix is capable of immobilizing or trapping extremely large amounts of water. More random aggregation or aggregation in which polymer-polymer interactions are favored over polymer-solvent reactions may be defined as protein coagulation. Empirically, there is unavoidable overlap in this terminology. At the macroscopic level, it may be difficult to differentiate between a highly solvated coagulum (or coagel) from a true protein gel.

Mechanisms of Protein Gelation

Formation of protein gel structures can occur under conditions which disrupt the native protein structure provided that the protein concentration, thermodynamic conditions and other conditions are optimal for the formation of the tertiary matrix. The most important food processing techniques relative to protein gelation involve divalent cations (calcium) and/or heat treatment.

0097–6156/81/0147–0131$05.00/0

Heat-induced Gelation. The mechanism suggested by Ferry in
1948 (3) is still the most generally accepted theoretical heat-
induced protein gelation mechanism. This two-step mechanism
involves: 1) an initiation step involving an unfolding or dissociation of the protein molecule followed by 2) an aggregation step
in which association and aggregation reactions occur resulting in
gel formation under appropriate conditions. For the formation of
the highly ordered gel matrix, it is imperative that the aggregation step proceed at a slower rate than the unfolding step (1,2).
In complex globular protein systems aggregation may occur more
randomly and simultaneously with the initial step. In these
instances the resultant gel (or coagel) would be composed of
strands in which spherical aggregated particles predominate as
described by the "string of beads" model of Tombs (4,5).
Because of the thickness and spatial arrangement of strands, this
type of gel would be characterized by higher opacity, lower elasticity and more syneresis (or weeping of solvent) than that of a
more highly ordered narrow stranded, or fibrous gel network. In a
general sense there are two basic types of heat-induced gel structures depending on the conditions involved. These gel types may
be generally termed: 1)thermo-set (or "set") or reversible and 2)
thermoplastic or irreversible gels. In thermo-set gelation, a sol
or progel condition can be demonstrated upon heating which is usually accompanied by increased viscosity. This progel "sets" to
form a gel upon cooling. This type of gel can usually be melted
to reform the progel upon subsequent heating suggesting that the
aggregation step is reversible (1,2). The progel state has been
observed in certain soybean protein gels (6,7). Gelatin gels,
depending on formation conditions, may also be characterized as
reversible gels (8). Thermoplastic or irreversible gels will
soften or shrink with subsequent heating, but melting or reversion
to the progel does not occur under practical conditions. It is
theoretically plausible that a given protein system may possess
the ability to form either type of gel depending upon the
formation conditions.

Calcium-induced Gelation. The mechanisms involved in the
gelation of proteins with cations are less defined than are those
suggested for heat-induced gelation. Several theories and models
have been proposed to describe the calcium-paracasein-phosphate
gel resulting from the enzymatic disruption of casein in the manufacture of cheese. It may be generally stated that calcium phosphate is involved in and is important to the gel network, but is
not solely responsible for the gelation phenomenon (9). The calcium-paracasein-phosphate gel system is characterized by low gel
strength and rapid syneresis. Other gelation processes which
involve addition of calcium (such as soybean tofu manufacture)
generally involve a combination of calcium addition and heating.
Extremely firm, highly expanded gel structures can be formed by
these combined processes (10).

Role of Crosslinking in the Gel Structure. In addition to effects of the size, shape and arrangement of the primary protein strands comprising the gel network, the characteristics of protein gels are affected by intra- and inter-strand crosslinking. This crosslinking combined with the fluidity of the immobilized solvent give gels their characteristic strength, elasticity and flow behavior. Degree of crosslinking must be optimal. Insufficient crosslinking or overdependence on crosslinking results in undesirable gel structure. Protein gels may be crosslinked by specific bonding at specific sites on the protein strands or by nonspecific bonding occurring along the protein strands. The nature and degree of crosslinking would vary with the type of protein and the gelation environment. The general types of crosslink bond, their characteristics and proposed role in protein gels have been summarized in Table I. Both covalent and noncovalent bonds are thought to be involved.

The irreversible nature of some protein gels suggests either a highly irreversible destruction in quaternary structure or the formation of covalent bonds. More reversible thermo-set gels would conceivably have less dependence on covalent bonding. Investigations of covalent bonding in heat-induced protein gels have focused primarily on disulfide bridging. Other types of covalent crosslinking suggested in gelatin gels (8) have not been investigated in more complex protein gel structures. Several possible reactions result in disulfide bridge formation. Heat treatment can result in cleavage of existing disulfide bonds structure or "activation" of buried sulfhydryl groups through unfolding of the protein. These newly formed or activated sulfhydryl groups can form new intermolecular disulfide bonds. In some protein systems, disulfide bridging may be imperative to the formation of a highly ordered gel structure (5). However, a high dependence upon disulfide bonding would tend to restrict solvent immobilization and result in more aggregated gel structures.

Definition of the role of disulfide bridging in protein gel systems has been primarily approached using thiol-reducing agents or sulfhydryl blocking agents. Reducing agents have been used to modify the gelation of proteins through modification of sulfhydryl interchange reactions (11,12,13,14,15,16). The action of these reagents, however, is concentration dependent, and care must be exercised in their use (14,15). Direct analysis of protein gels for sulfhydryl and disulfide groups has been used to demonstrate their involvement in milk gels (17).

Because of the dependence of sulfhydryl and disulfide reactions on heat treatment, it would be expected that their involvement in protein gelation would depend greatly on gel formation conditions. For example, Catsimpoolas and Meyer (7) have suggested that disulfides play only a minor role in the progel to gel transformation in soybean globulins. In more irreversible soy protein gelation processes, disulfide bridging may be involved in

the gel structure. Saio et al. (10,18,19) suggest that disulfide
bonding is important to gel structures formed from 11S soy
protein, but are not important to gel structures formed from the
7S fraction. The quantity of sulfhydryl and disulfide groups in
egg white or whey proteins would tend to support their involvement in heat-induced gel structure formation from these proteins.
     The noncovalent bonding involved in the crosslinking of pro-
tein gels include: 1) hydrogen bonding at specific sites, 2)
ionic bonding between charged amino acid side chains or as salt
bridges, and 3) hydrophobic and related attractions along protein
strands.
     Hydrogen bonding plays a major role in the increased viscos-
ity preceding the onset of gelation and in stabilizing the gel
structure. This type of crosslink allows for a more open orientation necessary for water immobilization. Hydrogen bonding may be
the most important type of crosslinking in reversible and gelatin
gels (8).
     Nonspecific hydrophobic attractions are important to the dis-
sociative-associative reactions which initate the gelation process. These attractions could also be involved in layering or
thickening of the gel network strands upon cooling which results
in improved strength and stability. This thickening of gel network strands also results in increased opacity (8).
     Ionic bonding may be of primary importance at the protein-
solvent interface and to solvent immobilization. However, indirect ionic effects on association-dissociation reactions in the
gelation process must be discussed. It is reasonable to assume
that effects of the ionic environment on protein gel solvation
would follow similar relationships as proposed for protein solubility. Gel solvation would therefore be expected to maximize
with increased salt addition due to decreased protein-protein
attractions and increased protein-solvent attractions. However,
gels formed at higher ionic strength would tend to exhibit more
protein-protein attraction as the ions compete with the protein
for solvent. A more aggregated nature of whey protein gels formed
at high ionic strength compared to gels formed at low ionic
strength has been clearly demonstrated by electron microscopy (2).

EXPERIMENTAL PROCEDURES

General Coagulation and Gelation Methodology

     Gelation and coagulation properties of proteins are complex
and difficult to interpret due to the extremely specific conditions required for gel or coagel formation. Moreover, gelation
and coagulation properties of proteins are complexly interrelated
to other protein functional properties. Gelation has been classified as a hydration, structural, textural and rheological property
of protein (20). An interrelationship between solubility, swelling, viscosity and gelation has been observed (21,22). Within

finite limits a high degree of protein solubility is necessary for
optimal protein gelation (23).

A wide variety of gel formation conditions and a wide variety
of measurement parameters have been used to describe these functional properties. Heat-induced gel formation techniques generally involve heating protein dispersions at appropriate concentrations in sealed test tubes (7,13,14,15,16,24,25), in sealed cans
(1,2,6,21,22), in specially designed gel tubes (23), or in sausage
casings (12). The following parameters may be important in characterizing protein gels (23): gel strength or hardness (resistance
to compression), adhesiveness (stickiness to other materials),
cohesiveness (stickiness to itself) and elasticity (ability to
regain original form after deformation under mild pressure).
These gel parameters can be obtained using texturometer or Instron
methodology. Other important parameters relate to shear and flow
(12) and water holding capacity. The single most measured parameter has been gel strength because of the relative ease in measurement. The following techniques can be used to evaluate gel
strength: 1) visual observation through comparison to gel standards (13,16), 2) viscosity with Helipath viscometers (6,7,21,22,
24), 3) penetration with specially designed probes for penetrometers (12) or Instron equipment (15) and 4) compression with a
variety of instruments such as jelly testers (23) or gelometers
(13).

The importance of the subjective or qualitative aspects of
protein gel systems should not be underestimated. For example,
the visual appearance of measurably strong protein gels may range
from that of an elastic translucent gel to that of a more opaque,
curd-like gel. Qualitative evaluation may be the most appropriate
means of differentiating between these protein gels at the macroscopic level. Electron microscopic analyses have been extremely
useful in evaluating protein gels (2,26).

Because of a lack of reliable methodology, there is sparse
information on the solvation and/or syneresis of gels formed from
globular proteins prepared under differing conditions. While
bound water can be obtained by a variety of techniques, the majority of the water held in the gel network is "free" water which is
more difficult to measure. The recently described methods based
on capillary suction potential (27) may have application to polysaccharide and gelatin gel systems. However the requirement of an
uniformly "set" gel surface in special glassware may limit the
usefulness of this technique in assessing water holding capacity
of irreversible gels. Data obtained from methods which involve
mechanical or physical release of the water from the gel are
difficult to interpret (15).

Preparation and Characterization of Whey Protein Gels

Procedures and conditions for whey protein concentrate (WPC)
gel formation have been described (14,15,16,25). Several commer-

cial whey protein preparations (Table II) have been evaluated for
their gel forming ability. Investigations of effects of salt and
reagent addition on gelation were performed using WPC-A (Enrpro
50, Stauffer Chemical Co., Rochester, MN). Aqueous protein suspensions were adjusted to pH 9.0 with sodium hydroxide and were
stirred for 30 min. The pH was then adjusted to the desired pH
with hydrochloric acid, and the suspensions were equilibrated at
22°C for 30 min. Dialyzed WPC was prepared by dialysis in an Amicon DC-2 hollow fiber apparatus (Amicon Corp., Lexington, MA).

For experiments investigating individual reagent effects on
WPC gelation (15,25), cysteine and calcium chloride were added at
levels ranging from 0 to 40 mM, and sodium chloride was added at
levels ranging from 0 to 0.5 M. In experiments investigating combined $CaCl_2$ and cysteine effects, reagent levels were selected
according to a central composite rotatable design (15). The dispersions were equilibrated with stirring at room temperature for
30 min following reagent addition.

For qualitative determination of gelation ability, 3 ml ali-
quots of the WPC dispersions were dispensed in screw capped test
tubes and heated in an oil bath at 100°C. Tubes were removed from
the bath at 30-s intervals and placed into an ice bath. Gel
strength was evaluated on a visual rating scale of 0 to 5.0 (13,
16). The time required at 100°C for the formation of a gel with a
rating of 4.0 or higher was reported as gel time.

Gels for quantitating gel characteristics were prepared in
screw capped centrifuge tubes (22 X 100 mm) by heating at 100°C
for 15 min (14,15,16,25). Gel strength was determined by a penetration technique using the Instron fitted with a disk probe (6.0
mm diameter) (15,16). Instrumental texture profile analysis on
gel sections (1.5 X 1.0 cm) was done on an Instron fitted with a
5.5 cm diameter probe operating at 2.0 cm/min to a 5.0 mm dis-
placement (14,15,25). Texture parameters of hardness, cohesiveness, and springiness were compared. Water-holding capacity of
gel sections was measured during compression on the Instron. The
"compressible water" was determined by increased weight of filter
paper (Whatman No. 3) due to water uptake during compression.
Regression equations of the form: $\hat{y} = \beta_0 + \beta_1 x + \beta_2 x^2 + \beta_3 x^3 + e$
were set up to measure effects of individual independent variables
(x), $CaCl_2$ or cysteine, on dependent variables (y) which were the
texture parameters. Regression coefficients (β_i) for combined
effects of $CaCl_2$ and cysteine were generated according to the mod-
el: $\hat{y} = \beta_0 + \beta_1 x_1 + \beta_2 x_2 + \beta_3 x_1^2 + \beta_4 x_2^2 + \beta_5 x_1 x_2$. Statistical
significance of the coefficients was determined by the "t test".

RESULTS AND DISCUSSION

Factors Affecting Gelation

For gelation/coagulation investigations to be useful measures
of protein functionality, it is essential that the conditions be

Table I. Proposed crosslink bonding of protein gel structures and their properties.

Type	Energy (Kcal/mole)	Interaction Distance	Groups Involved	Role in Gel Matrix
Covalent Bonding	80–90	1–2 A	–S–S–	Bridging; Ordering
Hydrogen Bonding	2–10	2–3 A	–NH$\cdots$O=C– –OH$\cdots$O=C–	Briding; Stabilizing
Hydrophobic and Related Interactions	1–3	3–5 A	Nonspecific	Strand thickening; Strengthening; Stabilizing
Ionic Bonding and Interactions	10–20	2–3 A	–NH$_3^+$; –COO$^-$, etc.	Solvent interactions; Salt linking

Table II. Method of preparation, composition and gelation characteristics of selected commercial whey protein concentrates (WPC)[a].

Protein System	Method of Preparation	Composition Ratio[b]		Gel Time[c]
WPC-A	Gel filtration	Protein/lactose	2.0	1.0
		Protein/fat	21.4	
		Protein/ash	5.2	
WPC-B	Electrodialysis	Protein/lactose	0.5	13.0
		Protein/fat	11.6	
		Protein/ash	2.6	
WPC-D	Ultrafiltration	Protein/lactose	1.2	30.0 +
		Protein/fat	16.6	
		Protein/ash	17.8	

[a]Adapted from Schmidt et al. (25).

[b]Ratio of protein/nonprotein component.

[c]Time (min) required at 100°C to form a visually strong gel.

defined and carefully controlled, and that the gelation conditions
be varied over a specified range. Variables to be considered
include: protein concentration, other protein and nonprotein
components, pH, ionic and/or reducing agents, and heat treatment
conditions.

Protein Concentration. For a given type of protein, a critical concentration is required for the formation of a gel and the
type of gel varies with the protein concentration. For example,
gelatin and polysaccharide solutions will form gels at relatively
low concentrations of the gelling material. Considerably higher
protein concentration is usually required for the gelation of
globular proteins.

As presented in Figure 1, a protein concentration of 7.5% or
higher was required for the formation of visually strong gels from
WPC heated at 100°C for 10 min at pH 7.0. The opacity of gels
formed in these experiments intensified with increased protein
concentration. Extending heating time of 5.0% protein dispersions
to 30 min resulted in enhanced gel strength.

Other Protein Components. Other protein components in complex food systems and in protein ingredient preparations may
interfere with or modify gelation reactions. Protein interaction
between whey protein and casein upon heating has a profound influence on the characteristics of the casein gel structure in cheese-
making. Similarly protein interactions are important to meat
structures. Protein-protein interaction between soy and meat proteins has also been demonstrated with heat treatment (28). While
concrete interaction data have not been collected on protein gels
formed from protein combinations, gelation properties of whey
protein/peanut flour blends have been investigated (25).

Nonprotein Components. More information is needed on the
effects of protein-lipid and protein-carbohydrate interactions on
protein gelation reactions. Effects of protein-lipid interaction
on soy protein gelation have been partially characterized (24).
The gelation of whey protein has also been shown to be affected by
the presence of lipids and the presence of lactose (29).

The data presented in Table II summarize differences in composition, preparation method and time required for gel formation
of 10% protein dispersions of WPC heated at 100°C. Differences in
gelling time could not be related entirely to compositional differences and appear to be related to other additional factors
(i.e., preparation technique, protein denaturation level, etc.)
(25). Differences in protein/lactose and in protein/fat ratio
could in part account for difference in gel time between WPC-A and
-B. However, WPC-D prepared by ultrafiltration had high protein
content relative to nonprotein components but did not form a gel.

Dialysis to minimize ash and lactose improved gelation properties of WPC systems with an ability to form gels but did not

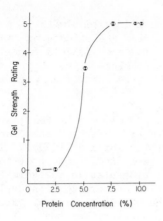

Journal of Food Science

Figure 1.   Effect of protein concentration on qualitative gel strength of WPC heated at 100°C for 10 min (16)

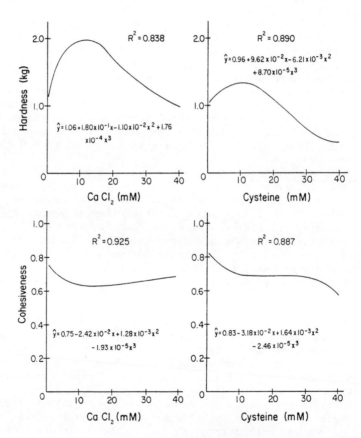

Journal of Agricultural and Food Chemistry

Figure 2.   Effect of CaCl$_2$ on cysteine addition to dialyzed WPC dispersions at 10.0% protein and pH 7.0 on hardness and cohesiveness of gels prepared by heating at 100°C for 15 min (15). Prediction equation $y = \beta_0 + \beta_1 x + \beta_2 x^2 + \beta_3 x^3$. Coefficients significant ($p < 0.01$).

appreciably affect the gelation of the WPC systems with poor gelling ability. Data in Table III summarize the effects of dialysis
on the gel characteristics of WPC-A. In general, gels formed
after dialysis of this WPC were more translucent, stronger, more
cohesive, less springy, more gummy and more chewy than were gels
formed from nondialyzed WPC (25).

   pH. Exposure to moderately high pH followed by readjustment
of the pH to neutrality has been shown to "activate" the protein
molecules thereby improving their gelling ability (13,30). This
activation step may be related to an unfolding of the protein
which would be important to the initiation step in the gelation
process. It may also involve activation of buried sulfhydryl
groups which would be important to the aggregation step. Prolonged exposure to extremely high pH, however, may suppress
aggregation.
   The pH of the heated protein dispersion would be expected to
have a profound effect on the gelation reactions. The effects of
pH on the gel characteristics of WPC dispersions heated at 100°C
are presented in Table IV. Quantitative decreases in gel strength
were apparent from penetration analyses as pH increased from pH
7.0 to 9.0 (16). However, with the exception of slightly more
browning at pH 9.0, the visual appearance of the gels formed in
this pH range was not appreciably different. Other researchers
have shown that WPC gels prepared at pH 8.5 were more elastic and
gelled at a lower temperature than gels formed at pH 6.0 (30).

   Ionic and/or Reducing Agents. Dialyzed WPC systems are more
responsive in terms of the effect of ionic or reducing agents on
gelation characteristics than are nondialyzed WPC systems and may
be useful as a model to study effects of these reagents (15,25).
Data presented in Figure 2 summarize $CaCl_2$ effects on WPC gel
characteristics. Maximum hardness occurred with addition of 5.0
to 20.0 mM added $CaCl_2$. Similar trends were noted with addition
of NaCl where gel strength maximized at 0.1 to 0.3 M NaCl (25).
Cohesiveness and springiness data tended to decrease slightly with
salt addition.
   As stated previously, the effects of sulfhydryl modifying
reagents (such as cysteine) are extremely concentration dependent
Gel hardness was predicted to maximize at 9.7 M cysteine while
cohesiveness decreased slightly with low levels of added cysteine
(Figure 2). Cysteine concentrations of 30 mM or higher virtually
destroyed the gelling capability of the WPC under the heating
conditions employed.
   Multiple regression and response surface analysis techniques
were employed to assess the effects of cysteine and $CaCl_2$ combinations on WPC gel characteristics (15). Contour plots summarizing
these data trends are presented in Figure 3. Significant interaction effects were observed between reagents with respect to hardness, cohesiveness and compressible water. At low levels of added

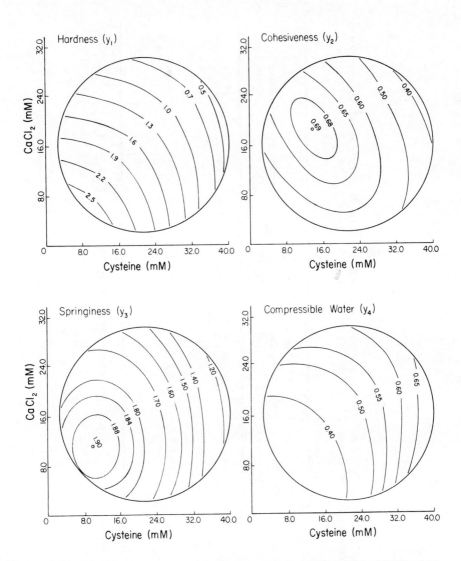

Journal of Agricultural and Food Chemistry

Figure 3. Response surface contour plots (15) for cysteine (x₁) and CaCl₂ (x₂) addition to 10% whey protein dispersions at pH 7.0 on hardness (y₁), cohesiveness (y₂), springiness (y₃), and compressible water (y₄) of gels prepared by heating at 100°C for 15 min

Table III. Dialysis effect on gel characteristics of 10% protein dispersions of whey protein concentrate (WPC) heated at 100°C for 15 min.

| Protein System | Visual | Gel Characteristics[a] | | | | |
		Hardness (Kg)	Cohesiveness	Springiness (mm)	Gumminess (Kg)	Chewiness (Kg-mm)
Non-Dialyzed	Firm; Opaque; Curdy	0.56	0.61	3.70	0.34	1.26
Dialyzed	Firm; Translucent	0.90	0.81	2.15	0.73	1.57

[a]Texture profile analyses on Instron as described by Schmidt et al. (25).

Table IV.   The pH effect on gel strength of whey protein concentrate (WPC) heated at 100°C for 15 min according to Schmidt et al. (<u>16</u>).

pH	Gel Strength	
	Qualitative[a]	Quantitative[b]
7.0	5.0	100.0
9.0	5.0	57.4
10.0	5.0	23.4
11.0	1.0	0.0

[a]Rating Scale: 0 = no gel, 5 = strong gel.

[b]Calculated from Instron penetration work data (as percentage of data collected on a similarly heated WPC control).

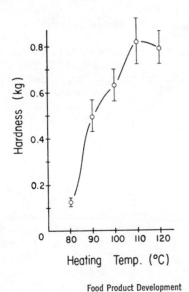

Heating   Temp. (°C)

Food Product Development

Figure 4.   Effect of heating temperature on gel strength of WPC formulated at 10.0% protein and heated for 15 min (14)

reagents, hardness values were slightly higher than was observed
with the reagents individually. Predicted data, in fact, suggest
that maximal hardness and minimal compressible water may be possible at a reagent combination outside the constraints of the experiments. Predicted gel hardness was at 2.6 mM cysteine and 2.5
mM $CaCl_2$. However, it may not be desirable to maximize hardness
and minimize compressible water in a protein gel structure. The
data presented suggest that ionic and reducing reagents act
synergistically in protein gelation.

Heat Treatment Conditions. Because of very finite process
constraints in the food processing industry, the heat treatment
required for gel formation is extremely critical when assessing
the applicability of a given protein to a specific food formulation. For example, replacement of myosin (which has relatively
low gel temperature requirement) in a hot dog formulation with a
protein preparation requiring extremely high temperature for
comparable gelation would not be desirable.

While gelation temperature is usually considered a characteristic property of a given protein system, the heating conditions
required for gel formation may be interrelated to all of the previously mentioned factors. It has been observed that WPC dispersions in 0.2 M NaCl will gel at 75°C while a temperature of 90°C
is required to gel WPC dispersions in distilled water (1). Heating time, at a specific temperature, required to form a protein
gel structure is generally considered to decrease with increased
protein concentration. Alteration of heat treatment conditions
affects the gel's macroscopic and microscopic structural attributes. This has been dramatically shown by Tombs (4) with
electromicroscopic evaluation of bovine serum albumin gels.

The effect of increased heating temperature on gel strength
of 10% WPC dispersions heated for 15 min is presented in Figure 4.
The dramatic increases in gel strength noted with increased heat
treatment were accompanied by slight decreases in cohesiveness and
an increase in opacity.

Gelation/Coagulation in Complex Food Systems

The structure of some foods can be directly perceived as a
gel (i.e., cheese, soybean tofu, gelatin, egg white, etc.). In
many semi solid (i.e., bread dough) or semi liquid (i.e., yogurt)
food products, however, the relationship between protein gelation/
coagulation properties and structure is not as easily defined.
Correlation of data obtained on the gelation/coagulation properties of proteins to their applicability in complex food systems
has only been sparsely investigated. Experiments are needed
relating these functional properties to the properties of model
and utility food systems with statistical modelling of the data.
This approach has been described by Hermansson & Akesson (21,22),
using the model meat system, and may prove invaluable in assessing
the applicability of new and existing protein sources to specific
food formulations.

CONCLUSIONS

Protein gelation and coagulation are necessary to the structure of many foods and are complexly related to the rheological properties of a variety of food systems. These physicochemical properties are interrelated with protein swelling, solubilization, association, dissociation and aggregation reactions. Gelation can be defined as a protein aggregation phenomenon in which polymer-polymer and polymer-solvent interaction are so balanced that a tertiary network or matrix is formed. This semi-elastic matrix is capable of immobilizing or entrapping large amounts of water in addition to other food components. Coagulation is protein aggregation in which polymer-polymer interactions are favored, resulting in a less elastic, less hydrated structure than that of a protein gel. Primary protein gelation and coagulation techniques used in food systems involve heat treatment and/or divalent cation (calcium) addition. Hydrogen bonding, disulfide bridging and hydrophobic attractions play major roles in crosslinking and stabilizing the structure of a protein gel or coagulum. While ionic crosslinking is important to the structure of calcium-induced gels, ionic attractions are primarily involved in solvent immobilization of heat-induced gels. The proposed mechanism of heat-induced gelation generally involves an initial unfolding or dissociation of the protein followed by association reactions which under appropriate thermodynamic conditions result in formation of the gel network structure. The gelation characteristics of whey protein systems can be improved by dialysis to remove interfering components. Altering protein concentration, pH and heating conditions or addition of salts and reducing agents modified the characteristics of whey protein gels. Synergistic effects of $CaCl_2$ and cysteine on gel characteristics have been suggested. The most investigated parameter characterizing a protein gel or coagulum has been firmness. Further characterization of protein gels will require evaluating other textural and water binding properties.

LITERATURE CITED

1. Hermansson, A.-M. *J. Texture Studies*, 1978, 9, 33.

2. Hermansson, A.-M. Aggregation and denaturation involved in gel formation. In "Functionality and Protein Structure." ACS symposium series 92. Ed. Pour-El, A. Am. Chemical Soc., Washington, DC, 1979, p. 81.

3. Ferry, J. D. *Adv. Protein Chem.*, 1948, 4, 1.

4. Tombs, M. P. Alterations to proteins during processing and the formation of structures. In "Proteins as Human Food." Ed. Lawrie, R. A. Butterworth Publ., 1970, p. 126.

5. Tombs, M. P. Faraday Discuss. Chem. Soc., 1974, 57, 158.

6. Circle, S. J.; Meyer, E. W.; Whitney, R. W. Cereal Chem., 1964, 41, 157.

7. Catsimpoolas, N.; Meyer, E. W. Cereal Chem., 1970, 47, 559.

8. Stainsby, G. The gelatin gel and the sol-gel transformation. In "The Science and Technology of Gelatin." Eds. Ward, A. G.; Courts, A. Acad. Press, New York, 1977, p. 179.

9. Ernstrom, C. A.; Wong, N. P. Milk-clotting enzymes and cheese chemistry. In "Fundamentals of Dairy Chemistry." Eds. Webb, B.; Johnson, A.; Alford, J. AVI Publ. Co., Westport, CT, 1974, p. 662.

10. Saio, L.; Kajikawa, M.; Watanabe, T. Agr. Biol. Chem. (Japan), 1971, 33, 1301.

11. Huggins, C.; Tapley, D. F.; Jenson, E. V. Nature, 1951, 167, 592.

12. Kalab, M.; Emmons, D. B. J. Dairy Sci., 1972, 55, 1225.

13. Pour-El, A.; Swenson, T. S. Cereal Chem., 1976, 53, 438.

14. Schmidt, R. H.; Illingworth, B. L. Food Prod. Develop., 1978, 12, 60.

15. Schmidt, R. H.; Illingworth, B. L.; Deng, J. C.; Cornell, J. A. J. Agr. Food Chem., 1979, 27, 529.

16. Schmidt, R. H.; Illingworth, B. L.; Ahmed, E. M. J. Food Sci., 1978, 43, 613.

17. Kalab, M. J. Dairy Sci., 1970, 53, 711.

18. Saio, K.; Sato, I.; Watanabe, T. J. Food Sci., 1974, 39, 777.

19. Saio, K.; Watanabe, T. J. Texture Studies, 1978, 9, 135.

20. Kinsella, J. E. Crit. Rev. Food Sci. Nutr., 1976, 7, 219.

21. Hermansson, A.-M.; Akesson, C. J. Food Sci., 1975, 40, 595.

22. Hermansson, A.-M.; Akesson, C. J. Food Sci., 1975, 40, 603.

23. Balmaceda, E. A.; Kim, M. K.; Franzen, R.; Mardones, B.; Lugay, J. C. Protein functionality methodology--standard tests. Presented at the 36th annual meeting of the Inst. of Food Technol., Anaheim, CA., 1976.

24. Catsimpoolas, N.; Meyer, E. W. Cereal Chem., 1971, 48, 159.

25. Schmidt, R. H.; Illingworth, B. L.; Ahmed, E. M.; Richter, R. L. J. Food Proc. Pres., 1979, 2, 111.

26. Kalab, M.; Emmons, D. B. Milchwissenschaft, 1974, 29, 585.

27. Lewicki, P. P.; Busk, G. C.; Labuza, T. P. J. Colloidal & Interface Sci., 1978, 64, 501.

28. Peng, I. C.; Allen, C. E.; Dayton, W. R.; Quass, D. W. Protein-protein interaction between soybean 11S protein and skeletal muscle myosin. Presented at the 36th annual meeting of the Inst. of Food Technol., St. Louis, MO, 1979.

29. Sternberg, M.; Chiang, J. P.; Eberts, N. J. J. Dairy Sci., 1976, 59, 1042.

30. Ishino, K.; Okamoto, S. Cereal Chem., 1975, 52, 9.

31. Haggert, T. O. R. N.Z. J. Dairy Sci. and Technol., 1976, 11, 244.

RECEIVED September 5, 1980.

Whippability and Aeration

JOHN P. CHERRY

Southern Regional Research Center, Agricultural Research,
Science and Education Administration, U.S. Department of Agriculture,
P.O. Box 19687, New Orleans, LA 70179

KAY H. McWATTERS

Department of Food Science, University of Georgia College of Agriculture
Experiment Station, Experiment, GA 30212

A half century ago, researchers described foams as disperse structures that contain a colloidal liquid, such as a protein solution, as the dispersion medium and a gas or air as the disperse phase (1, 2, 3, 4). The factors principally involved in foam formation were surface tension, viscosity, and the character of the film that formed at the surface of the liquid. The foam was stable when low tension and high viscosity occurred at the surface of the colloidal solution, forming a tough or amorphous solid surface film. This sequence of events, described as the adsorption effect, depends upon the concentration of the colloid and its ability to denature and precipitate at the interface. Physicochemical changes of proteins at the interface are important properties that contribute to surface tension and viscosity of stable foams (5).

Theories on the Role of Proteins in Foams

Cumper and Alexander (6) and Cumper (7) explained that during foam formation, a monolayer of surface denatured protein surrounded by liquid is rapidly adsorbed at the interface of the colloidal mixture, trapping air and forming bubbles (Table I, II; Figure 1). Adsorption continually occurs around the bubbles to replace protein in areas of the interface where coagulation or stretching of the film is occurring. The actual bubble size in the foam depends upon the rate of protein adsorption as well as upon the ease of film rupture. The protein films on adjacent bubbles come in contact and trap the liquid, preventing it from flowing freely. This restriction is governed by the viscosity of the colloidal solution. The polypeptides of denatured proteins situate to positions where their hydrophobic side chains are directed outward toward each other. Because liquid

0097–6156/81/0147–0149$07.00/0

Table I. Sequence of events occurring during foam
          formation and coalescence.

1. Denaturation: uncoiling of protein polypeptides.

2. Adsorption: formation of a monolayer or film of
        denatured protein at the surface of the colloidal
        solution.

3. Entrapment: surrounding of gas at the interface
        by the film and formation of bubbles.

4. Repair: continued adsorption or formation of a
        second monolayer around the bubbles to replace
        coagulated regions of the film.

5. Contact: protein films of adjacent bubbles come
        in contact and prevent flow of the liquid.

6. Coagulation: interacting forces between polypeptides
        increase causing protein aggregation and weakening
        of the surface film followed by bursting of the
        bubble; weakening of the film also occurs when the
        Repair step ceases because of a deficiency of
        denatured protein.

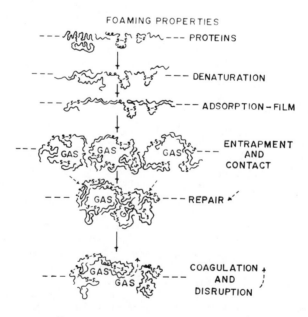

Figure 1.   Diagram of foam properties

Table II.   Factors affecting foaming properties of proteins.

1.   Concentration, Surface, and Bulk Viscosity: solubility,
        diffusion rate, interaction in the disperse phase, and
        increase of bulk with such substances as sucrose.

2.   Structure: disordered, flexible versus globular, rigid;
        availability of hydrophobic and hydrophilic groups.

3.   Denaturation: ease of polypeptide unfolding.

4.   Electrical Double Layer: repulsion affected by counter
        ions such as salts in solution; availability of
        hydrophobic and hydrophilic groups.

5.   Marongoni Effect: ability to concentrate rapidly at a
        stress point in the film.

6.   pH: maximum near the isoelectric point; extremes,
        dissociate polypeptides.

7.   Temperature: dissociate polypeptides.

8.   Denaturants: improve availability and interaction of
        polypeptides; e.g., thio-reducing reagents.

9.   Complementary Surfactants: other proteins, poly-
        saccharides; not lipids.

10.   Coagulation: irreversible aggregations.

and air are still moving through the foam, salt and hydrogen bonds form between the polypeptides of the proteins to interact with one another and together with interactions between nonpolar portions of the proteins, resulting in small aggregates being formed; that is, the proteins coagulate into aggregates and the bubbles burst.

The polar groups in the proteins cause the polypeptides to uncoil at the interface (Table I, II; Figure 1); the hydrophobic portions keep the film coherent. If these polar groups are inactivated, either by changing the external environment of the molecules or through internal salt formation (which is enhanced by agitation of the film), the spreading tendency decreases and the cohesion between the polypeptide chains increases. Any reduction in the degree of hydration of the polar groups facilitates the coagulation. With the majority of proteins, this process is irreversible and results in their coagulation. In addition to breaking of bubbles in the foam, it also results in a reduction in the quantity of proteins available for foam stabilization.

Certain basic mechanical and chemical characteristics common to all foaming liquids were noted and differentiated them from nonfoaming substances (8). Persistent foams were noted to arise only with solutes that lower the surface tension in a solution, or are highly surface-active substances (for example, soaps, synthetic detergents, proteins, and so forth). Foams remain stable as long as the liquid maintaining the three-dimensional structure, or lamellae, between the bubbles did not drain and allow the entrapped gas to diffuse out. To show the special rheological property of film elasticity, the lamellae must have a restoring force strong enough to counterbalance local thinning caused by a loss of liquid or applied stress. The surface elasticity of the solution is an especially important property because it is at this point that the air is entrapped. The phenomenon of surface transport, by which a spreading monolayer drags with it significant quantities of the underlying solution and rapidly repairs thinning spots in a lamella, was suggested to occur in quality foaming solutions. The lifetime of foams can be greatly extended by increasing the viscosity of the liquid or its surface layer; that is, enhanced viscosity retards drainage of liquid from between the bubbles.

Electric repulsion between the ionic double-layers that are formed by adsorption on two sides of the lamellae was suggested to contribute to the stability of foams. Foaming may be either enhanced or reduced by addition of a second surface-active agent and depends upon the ability of the two agents to pack together in a mixed adsorption layer. Such surface complexes seem to depend on the convenient packing of the nonionic agents between the chains of the ionic ones, thus causing condensation of a monolayer that would otherwise be expanded due to electrostatic repulsion between the head groups.

β-casein, a structurally disordered, flexible protein, adsorbs rapidly at the air-water interface and gives good foamability (9). High-ordered globular proteins such as lysozyme and bovine serum albumin were difficult to surface-denature and as a result had poor foamability. Although the globulins arrive at the interfacial region at a similar rate as that of β-casein, the proportion of their available amino acids in the interface is less than that of the flexible protein. Because of their rigidity, the partial unfolding and rearrangement of the globulins in the interface is relatively slow compared to β-casein. Consequently, air cells formed during shaking are not stabilized effectively, and the rate of increase of foam volume with time is relatively low. Prior denaturation of the globulins by heat or extremes of pH under conditions where no precipitation occurs enhances foamability of these proteins (10). On the other hand, foams formed with solutions containing globular proteins are more stable than those of flexible β-casein because interactions of globulins are more rigid.

Horiuchi et al. (11) confirmed the earlier work (6, 7) showing that foam stability occurred when the hydrophobic regions of proteins became situated at the interface, causing these molecules to resist migration into the water phase. Mercaptoethanol (a thio-reducing agent that dissociates intrapeptide disulfide bonds), partial proteolysis, or heat treatment of soybean globulins caused unfolding of their polypeptides at the interface; this facilitated hydrophobic interactions, increased film thickness and viscosity, and improved foam stability. Cleaving and blocking disulfide bonds of wheat proteins (gluten, gliaden) significantly decreases bubble size and coalescence rate of foams (12).

Kinsella (13, 14) summarized present thinking on foam formation of protein solutions. When an aqueous suspension of protein ingredient (for example, flour, concentrate, or isolate) is agitated by whipping or aeration processes, it will encapsulate air into droplets or bubbles that are surrounded by a liquid film. The film consists of denatured protein that lowers the interfacial tension between air and water, facilitating deformation of the liquid and expansion against its surface tension. The proteins in the liquid film should 1) be soluble in the aqueous phase, 2) readily concentrate at the liquid-air interface, and 3) denature to form cohesive layers possessing sufficient viscosity and mechanical strength to prevent rupture and coalescence of the droplet. That is, the polypeptides of the denatured proteins in the liquid film should exhibit a balance between their ability to associate and form a film and their ability to dissociate, resulting in foam instability.

Experimental Procedures

Many methods have been used to produce and characterize protein foams. The foaming characteristics of proteins are markedly influenced by conditions of preparation, measurement, and so forth (13-18). Because of the variety of methods employed, it is difficult to compare data from different sources.

Three dynamic procedures for determining foaming capacity of proteins include whipping, shaking, and sparging (7, 16). Differences between these methods include 1) amount of protein required for foam formation (3-40% for whipping, approximately 1% for shaking, and 0.01-2% for gas sparging) (16, 19, 20); and 2) way the foams are formed (shear forces are important in whipping and shaking, but not in sparging).

Whipping produces foams that can be measured by the increase in foam volume, specific gravity, and viscosity (15, 17, 19). Rapid shaking of a horizontal graduated cylinder containing a protein solution produces a foam that can be measured by its volume (9, 16). Foaming capacity of sparged foams is measured by the ratio of the volume of gas in foam to the volume of gas sparged, or by the maximum volume of foam divided by the gas flow rate (7, 10, 20, 21).

The stability of foams is usually measured by the volume of liquid drained from a foam during a specific time at room temperature (10, 15, 17) or by a decrease in foam volume over time. Methods employed to measure foam stability include the rate of fall of a perforated weight through a column of foam (20, 21), the penetration of a penetrometer cone (17), or the ability to support a series of specific weights (22).

Waniska and Kinsella (18) developed a specially designed small-scale foaming apparatus to determine dynamic and static foaming properties of proteins. Foam was produced by sparging nitrogen at a known rate through a dilute protein solution maintained at known temperature in a water-jacketed column. Foaming properties readily examined were capacity, strength, and stability. The apparatus permitted many of the variables affecting foaming properties of proteins (that is, pH, temperature, ions, carbohydrates, surfactants) to be controlled while quantitatively determining foaming properties of small quantities of proteins.

For the studies presented in this chapter, samples of peanut and cottonseed meal suspensions were evaluated for foam capacity, stability, and viscosity measurements as described by Cherry and coworkers (23, 24, 25). Vegetable protein suspensions at the appropriate concentration and pH were whipped in a Waring-type blender. After blending, the whipped products were transferred to a graduated cyclinder. Milliliters of foam were recorded immediately and at various time intervals to determine capacity and stability. A Brookfield viscometer and

Helipath stand equipped with a T-C spindle operated at a set rpm was used to measure viscosity in centipoises of freshly prepared foams after they were allowed to stand for various times.

Results and Discussion

Foaming properties related to pH. Quantity of soluble proteins of glandless cottonseed flour declined in suspensions at pH values between 1.5 (which contained acid-dissociated proteins) and 4.5 (isoelectric pH of cottonseed proteins), and then gradually increased as the pH was raised to neutrality (water-soluble proteins; Figure 2; 24, 25). Amounts of soluble storage globulins increased as the pH was raised to 11.5. These changes are shown qualitatively by polyacrylamide disc-gel electrophoretic patterns in Figure 3. Protein quantities of the insoluble fractions at various pH values were the inverse of those of the soluble portions (Figure 4).

Carbohydrate content increased gradually in soluble fractions of flour suspensions as their pH values were raised from 1.5 to 11.5 (Figure 2). Ash levels were highest in soluble fractions at pH values between 3.5 and 5.5, and fiber was practically negligible in all of these preparations. In the insoluble fractions, carbohydrate content was increased to the highest content when the pH of the suspension was adjusted to values between 5.5 and 6.5, and remained near these high levels to pH 11.5 (Figure 4). Ash percentages were highest in insoluble fractions between pH 1.5 and 5.5, and 7.5 and 8.5. Little change was noted in fiber content of insoluble fractions between pH 1.5 and 8.5; these values were signficantly greater than those of the soluble fractions. A decline in quantity of fiber was noted at the extreme alkaline pH values.

It was increasingly difficult to form foams of high stability with suspensions of cottonseed flour as the pH was raised from 1.5 to 3.5 (Figures 2 and 4). Values were similar for suspensions at pH 3.5 and 4.5. Between pH 4.5 and 6.5, foam capacity and stability declined to the lowest values noted in the experiment. As the pH was raised to 11.5, both of these functional properties increased to values similar to those at the extreme acid pH range. Thus, optimum foaming properties were noted when the acid pH-dissociated proteins and major storage globulins were most soluble (Figures 2, 3, and 4). The leveling off of foam capacity and stability at pH 3.5 and 4.5 may be related to the increase in soluble ash (Figure 2) or an increase in select proteins that are insoluble in the suspension (Figures 3 and 4) at these pH values.

Foam viscosity of freshly made foams, and those allowed to stand for 60 min, was highest for suspensions in the pH range of 3.5 to 5.5 (Figure 5); maximum foam viscosities after one and 60 min were at pH 3.5 and 5.5, respectively.

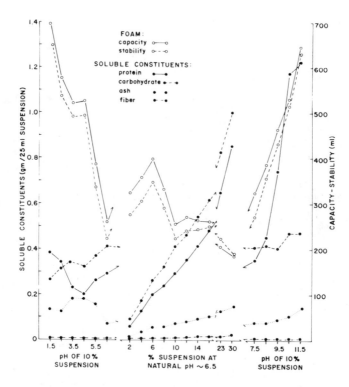

Figure 2.   Foam properties and content of soluble constituents related to pH and percentage of glandless cottonseed flour in aqueous suspensions

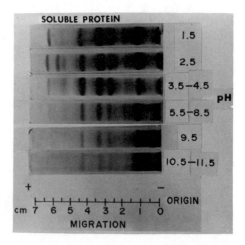

Figure 3.   Gel electrophoretic properties of glandless cottonseed proteins that are soluble at various suspension pH values
(25)

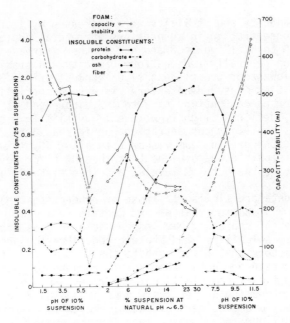

Figure 4.   Foam properties and content of insoluble constituents related to pH and percentage of glandless cottonseed flour in aqueous suspensions

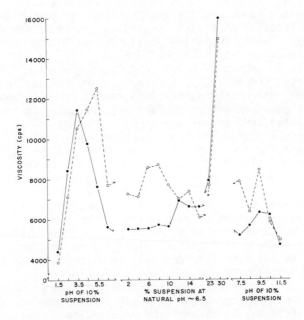

Figure 5.   Foam viscosity and stability properties of glandless cottonseed flour suspensions at various pH values.   Foam viscosity: (● — ●) 1 min; (○ – – – ○) 60 min.

Foam capacity and stability were increased by increasing the flour concentration in suspensions at natural pH (about pH 6.5) from 2% to 6% (Figures 2 and 4); further increases in flour concentration generally reduced foam capacity and stability which were lowest at the highest flour concentration (30%). Foam viscosities measured after one min were essentially the same at flour concentrations of 2 to 10%, increased at the 12% flour level, declined slightly at the 14 and 16% levels, then increased as the flour concentration was increased to 23 and 30% (Figure 5). Changes in foam viscosities after 60 min were more variable; viscosities were highest at the 30% flour level, intermediate at the 6 to 8% levels, and lowest at the 16% flour level.

These data are similar to those of other researchers (13, 15, 20, 26, 27) showing that pH and protein (especially large molecular weight storage globulins and acid pH-dissociated polypeptides) solubility, and protein concentration are important factors that contribute to the foamability of vegetable protein products. Nonprotein components that occur naturally in flour, especially carbohydrates and ash, many also contribute to foaming properties. A whipped product containing mainly soluble protein, carbohydrates, and salts was prepared from soybean flour and had excellent foaming properties (28). This material was successfully substituted for egg white in recipes for meringues, divinity-type candies, and souffles.

Empirical multiple linear regression models were developed to describe the foam capacity and stability data of Figures 2 and 4 as a function of pH and suspension concentration (Tables III and IV). These statistical analyses and foaming procedures were modeled after data published earlier (23, 24, 29, 30, 31). The multiple R^2 values of 0.9601 and 0.9563 for foam capacity and stability, respectively, were very high, indicating that approximately 96% of the variability contributing to both of these functional properties of foam was accounted for by the seven variables used in the equation.

Multiple linear regression equations were also developed for foam capacity and stability based on pH and the data on composition of soluble and insoluble fractions in the suspensions summarized in Figures 2 and 4 (Tables V and VI). Multiple R^2 values of 0.9346 and 0.9280 were obtained for these equations of capacity and stability, respectively. The relative importance of each respective partial regression coefficient was determined by comparison of β values (32). These evaluations indicate that the most important variables in the two models for foam capacity and stability are soluble protein, soluble and insoluble carbohydrate and ash, and insoluble fiber.

The experimental data in Figures 2 and 4 were used in multiple regression equations to predict foam capacity and stability of 2% to 30% suspensions adjusted to pH values of 1.5 to 11.5 (Figures 6 and 7). Observed and predicted data of the

Table III. Empirical multiple linear regression model[1] describing foaming capacity as a function of pH and suspension concentration.

Variable Y	Component	Regression Coefficient
	Intercept	608.360
X_1	pH	-53.484
X_2	Suspension concentration	41.021
X_3	pH squared	3.175
X_4	Suspension concentration squared	0.647
X_5	pH times concentration	17.016
X_6	pH squared times concentration	1.476
X_7	pH squared times concentration squared	-0.016

[1]Multiple R^2 = 0.9601.

Table IV. Empirical multiple linear regression model[1] describing foam stability as a function of pH and suspension concentration.

Variable Y	Component	Regression Coefficient
	Intercept	474.730
X_1	pH	-27.477
X_2	Suspension concentration	51.606
X_3	pH squared	0.740
X_4	Suspension concentration squared	0.785
X_5	pH times concentration	-20.412
X_6	pH squared times concentration	1.836
X_7	pH squared times concentration squared	-0.019

[1]Multiple R^2 = 0.9563.

Table V. Multiple linear regression analysis[1] of foam capacity
on compo⁻˙ːion of glandless cottonseed flour.

Variable Y	Component	Beta value
X_1	Soluble protein	0.906
X_2	Insoluble protein	-0.069
X_3	Soluble carbohydrate	-2.353
X_4	Insoluble carbohydrate	0.709
X_5	Soluble ash	0.624
X_6	Insoluble ash	0.219
X_7	Soluble fiber	-
X_8	Insoluble fiber	0.349

[1]Multiple R^2 = 0.9346.

Table VI. Multiple linear regression analysis[1] of foam
stability on composition of glandless cottonseed
flour.

Variable Y	Component	Beta value
X_1	Soluble protein	0.807
X_2	Insoluble protein	-0.474
X_3	Soluble carbohydrate	-2.354
X_4	Insoluble carbohydrate	1.030
X_5	Soluble ash	0.718
X_6	Insoluble ash	0.271
X_7	Soluble fiber	0.032
X_8	Insoluble fiber	0.480

[1]Multiple R^2 = 0.9280.

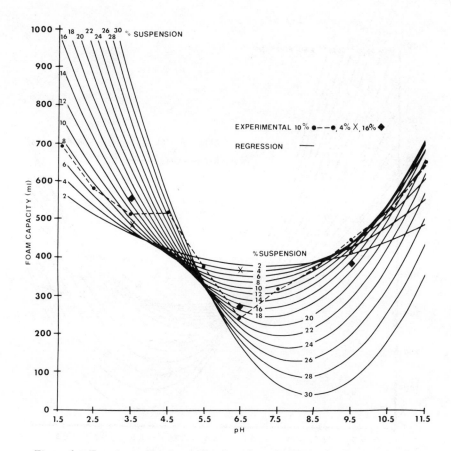

Figure 6. Experimentally observed and mathematically simulated regression lines of foam capacity of different percentages of glandless cottonseed flour in suspensions at various pH values. Experimental 4%, 10%, and 16% suspensions were run at pH 3.5, 6.5, and 9.5 to test the reliability of the multiple linear regression analysis. Quantitative data used in this analysis are in Figures 2 and 4.

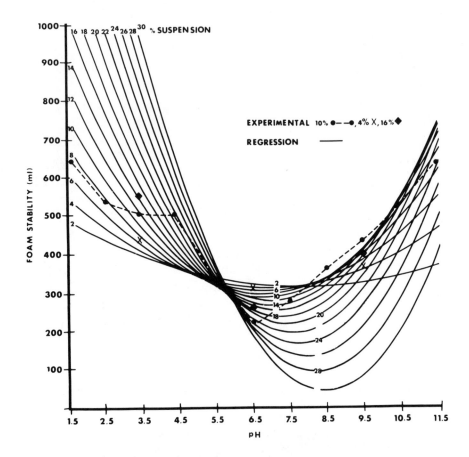

Figure 7. Experimentally observed and mathematically simulated regression lines of foam stability at different percentages of glandless cottonseed flour in suspensions at various pH values. See Figure 6 for further explanation of the data.

foam properties were similar for the 10% suspension at most of
the pH values and for the 2% to 30% suspensions at pH 6.5.
Differences between the predicted and observed curves were
mainly noted at the extreme acid pH values and at the
isoelectric pH 4.5. Observed and predicted foam capacities and
stabilities of 4%, 10%, and 16% suspensions at pH 3.5, 6.5, and
9.5, serving as checks, were comparable.
   Multiple regression analysis is a useful statistical tool
for the prediction of the effect of pH, suspension percentage,
and composition of soluble and insoluble fractions of oilseed
vegetable protein products on foam properties. Similar studies
were completed with emulsion properties of cottonseed and peanut
seed protein products (23, 24, 29, 30, 31). As observed with
the emulsion statistical studies, these regression equations are
not optimal, and predicted values outside the range of the
experimental data should be used only with caution. Extension
of these studies to include nonlinear (curvilinear) multiple
regression equations have proven useful in studies on the
functionality of peanut seed products (33).

   Foam properties related to salt. The addition of sodium
chloride to soybean protein suspensions caused them to form
high-capacity, low-stability foams (13). It was suggested that
foam capacity increased because salt improved protein solubility
at the interface of the colloidal suspension during foam
formation, but retarded the partial denaturation of the surface
polypeptides of proteins that are necessary for protein-protein
interaction and stability.
   Adjusting the pH of defatted peanut meal suspensions
containing 0, 0.1, and 1.0 N NaCl before whipping significantly
affected foam capacity and stability (Figure 8). The largest
increases in foam capacity occurred at pH 1.5. At this pH,
increasing the salt content significantly reduced the percentage
of soluble protein, did not alter foam capacity, but did reduce
foam stability. At pH 4.0, percentages of soluble protein were
among the lowest observed in this study, but foam capacity of
both water and salt suspensions was better than that at the
natural pH of 6.7; the 1.0 N NaCl suspension had the highest
capacity and was the most unstable.
   The smallest increases in foam capacity occurred at pH 6.7.
Percentages of protein in the soluble fractions varied at this
pH. Foam capacity and stability improved and decreased,
respectively, at this pH as the salt content of the suspension
was increased. The two-step pH adjustment (23) of 6.7 to 4.0 to
6.7 improved foam capacity (especially in the suspension
containing 1.0 N NaCl over that of the unadjusted pH 6.7
suspensions despite variation in percentage of protein in
soluble fractions. As foam capacity was improved in the
two-step adjusted pH suspensions, foam stability declined.

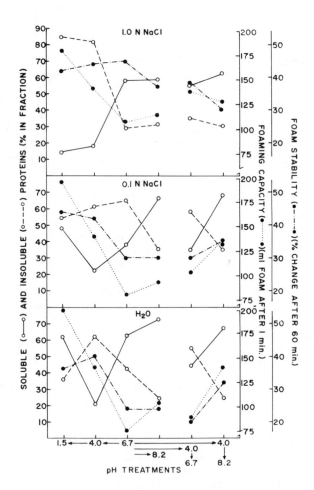

*Figure 8.   Effect of pH and salt concentration on the foam and protein solubility
properties of peanut meal suspensions (25)*

At pH 8.2, percentages of soluble protein in water and salt
suspensions were among the highest observed in the study, yet
foam capacities were not as great as those noted at pH 1.5 and
4.0. Suspensions with and without salt that were adjusted from
pH 6.7 to 4.0 to 8.2 contained high percentages of soluble
protein and exhibited foaming properties similar to those of
preparations at pH 4.0. A slight inhibition of foam formation
was noted with suspensions subjected to the two-step pH
adjustment to 8.2 as salt content was increased. The stability
of these foams was better than that of foams made with acid pH
suspensions.

The improved foam capacity of defatted peanut meal
suspensions in the pH 1.5 suspensions is probably related to the
dissociated proteins (regions 1.0-2.0 and 4.5-6.0) that are
distinguished on polyacrylamide gels (Figure 9); this improvement is especially noted in the water and 0.1 $\underline{N}$ NaCl suspensions. However, these bands are not clearly shown in the gels
of the high-salt-containing suspension. At pH 4.0, increasing
salt content of the suspension raised the number of bands that
could be distinguished in the gels; foam capacity was improved
at this pH when salt was added to the suspensions. At the
higher pH values (adjusted and not adjusted), the gel patterns
did not distinguish any differences among suspensions.

The foams produced at pH 1.5 were relatively stiff and
peaked with small air bubbles. Those produced at pH 6.7 and 8.2
were frothy, lacked stiffness, and had large bubbles. The pH of
a suspension, with or without salt, is an important
consideration in determining the type, capacity, and stability
of foam produced with whipping. Adjusting the pH to levels
above and below the natural pH 6.7 seems to be an effective
means of improving the foam properties of peanut seed meal
suspensions. Although not clearly shown by gel electrophoretic techniques, this adjustment of the pH evidently is
causing structural changes in the proteins, enabling them to
denature more readily and interact to improve foam properties.

Foam properties related to heat. Peanut kernels were moist
heated in a temperature-controlled retort at 50, 75, and 100°C
for 15-min intervals ranging from 15 to 90 min (30). Quantities
of water-extractable proteins decreased as heating time
increased from 15 to 90 min at all three temperatures (Figure
10). At 50°C, reduction in protein extractability was greater
after 15 min of heating than at 75 or 100°C. Little difference
in solubility was observed among temperatures at 30 min. At 45,
60, and 75 min heating times, decreases in protein solubility
became more notable with increasing temperature. Levels of
soluble protein were especially reduced when heating time at
100°C was extended beyond 60 min. The samples heated at 75°C
for 90 min showed an increase in protein solubility to levels

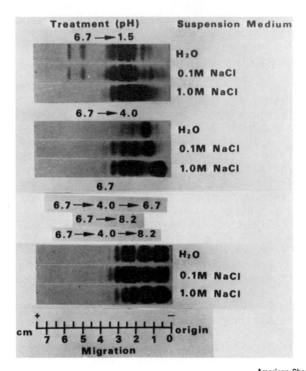

American Chemical Society

Figure 9.   Gel electrophoretic properties of peanut meal proteins that are soluble in suspensions that contain various levels of salt and are at different pH values (25)

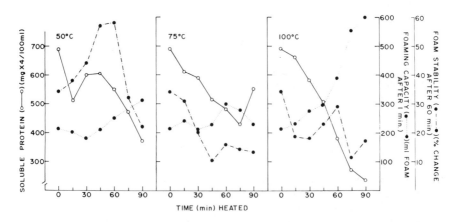

Figure 10.   Effect of moist heat on the foam and protein solubility properties of peanut meal suspensions

comparable with those treated at 50°C for 60 min, and 100°C for 30 min.

Foam capacities of meals from peanut seeds moist heated at 50°C for 0 to 30 min were similar, then increased as the heating time was extended beyond 30 min (Figure 10). The stability of these foams decreased, showing an increase in percentage of change to about 60% between 0 and 60 min (Figure 10); kernels heated for 75 and 90 min formed relatively stable foams (20 to 30% change). At 75°C, samples heated for 30 to 60 min showed a similar increase in capacity as those heated at 50°C; a slight decrease in capacity followed to 90 min. Foams from kernels heated at 75°C for 45 to 90 min had less percentage change in volume after 60 min than those heated for 0 to 45 min. The samples heated at 100°C for 15 to 90 min produced the largest increase in foam capacity. These foams had variable stability, with the least percentage change occurring at 75 min and the most at 0 min.

Polyacrylamide electrophoretic disc gels showed no detectable changes in water-extracted protein of peanut kernels moist-heated at 50 and 75°C up to 90 min (Figure 11). At 100°C, major structural changes, including electrical charge, conformation, and size, occurred in nonarachin proteins (region 1.5-3.5 cm) from samples heated for 30 to 90 min. The major components of the storage globulin, arachin, having mobilities in the gels to region 0.5-1.5 cm, were unchanged by the heat treatment of 100°C. The arachin was gradually altered to polypeptide subunits and aggregates by external heating for 105 to 180 min at 100°C (30).

Although gel electrophoretic and solubility techniques revealed few major changes in proteins of meals from peanut kernels moist-heated at 50°C and 75°C for time intervals of 15 to 90 min, variations in the extractability of these components suggest that structural changes are occurring. Structural changes in proteins of meals from seed heated at 100°C for various time intervals are readily shown by these techniques. Foam capacity of heated meals containing these denatured proteins is vastly improved.

The stability of these foams is improved only with samples heated for certain time intervals, for example, 0 to 45 min at 50 and 75°C, and 0 to 30 min and 60 to 75 min at 100°C. At 100°C and 60 to 75 min, major changes in proteins are occurring, as shown on the gel electrophoretic patterns.

Ingredients containing protein that had been heat denatured, but not to the degree that caused them to precipitate, enhance foamability (13). For example, controlled heat treatment of soybean globulins caused unfolding of their polypeptides at the interface of the foaming solution, facilitating hydrophobic interactions, increasing film thickness and viscosity around the bubbles, and improving foam stability. Similar observations have been noted with whey protein (15).

Proteins of egg white denature more rapidly than those of whey protein concentrate (13, 34). However, isolated β-lactoglobulin from the whey concentrate was more susceptible to surface denaturation than egg white ovalbumin. These data suggest that whey contains substances that protect the proteins from surface denaturation and may account for the lower stability of whey protein concentrate foams than those of egg white protein. A balance between the disaggregation effect of select pH values and the tendency toward greater aggregation of proteins at higher heating temperatures were correlated closely with maximum foam stability (13, 15).

Foam properties related to protein acylation. Succinic anhydride reacts with the Σ-amino group of lysine in protein, causing formation of negatively charged N-acylated residues. Other reaction products of protein include o-succinyltyrosine and succinic derivatives of aliphatic hydroxy and sulfhydryl groups of amino acids. Succinylation causes unfolding, expansion, and dissociation of protein polypeptides. An increase in negative charge is noted for acylated proteins. Modifications of proteins induced by acylation greatly affect the performance of oilseed ingredients when used in model functionality tests (35-46).

The solubility of proteins in suspensions containing liquid cyclone processed cottonseed flour that had been treated with 40% and 80% succinic anhydride (compared to 0% and 10%) was increased significantly in suspensions adjusted to pH 6.5 and 8.0 (Figure 12). Gel electrophoretic patterns showed an increase in proteins that migrated to region 1.0-3.0 cm in the suspensions containing 40% and 80% succinylated flour (Figure 13). This increase in protein in region 1.0 to 3.0 cm of the gels was also noted in suspensions at pH 2.5 and 5.0. Fewer bands were present in the lower half of these gel patterns compared to those of flours with 0% and 10% succinic anhydride. These data suggest an increase in size of proteins in succinylated flours; changes in charge caused by conformational alterations cannot be discounted. Some improvement in protein solubility was noted in pH 5.0 suspensions containing flours treated with high amounts of succinic anhydride. Treating the proteins with 10%, and especially 40%, succinic anhydride (compared to those of 0% and 80%) increased percentage of foam at pH 1.5. Little change was noted at pH 5.0 for all suspensions included in this study. At pH 6.5, 10% succinic anhydride treatment lowered foam increase, whereas 40% and 80% experimental conditions improved foam capacity. Only the 40% succinylation treatment improved foam capacity over that not treated, at pH 8.0.

Compared to the nontreated sample, all three succinylation treatments of flour improved foam stability at pH 6.5. At pH

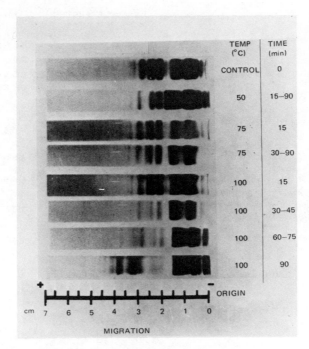

Figure 11. Gel electrophoretic properties of peanut meal proteins that are soluble in suspensions of meal from peanut seeds that were moist heated at various temperatures for different time intervals

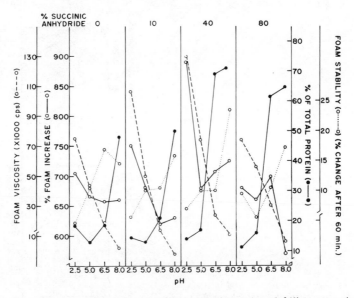

Figure 12. Effect of succinylation on the foam and protein solubility properties of liquid cyclone processed cottonseed flour in suspensions at various pH values

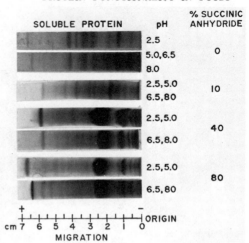

Figure 13. Gel electrophoretic properties of soluble proteins in suspensions at
various pH values that contain succinylated liquid cyclone processed cottonseed
flour

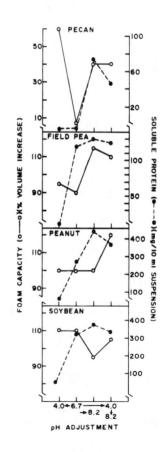

Figure 14. Foam capacity and protein
solubility properties of defatted soybean,
peanut, field pea, and pecan seed flour
suspensions at various pH values (47)

8.0, only the suspension containing 40% succinic anhydride-
treated flour showed a significant decline in foam stability.
In most cases, foam stabilities of suspensions at pH 2.5 and 5.0
indicated that 10% and 40% succinylated flours were not
different from the nontreated sample. The pH 5.0 suspension
containing flour that was treated with 80% succinic anhydride
formed very stable foams.

Foams of pH 2.5 suspensions containing 10% and 40%
succinylated flours were significantly more viscous than those
treated at the 0% and 80% levels. All other foams had similar
viscosities at similar pH treatments.

These data demonstrate that changes in foam properties of
liquid cyclone processed cottonseed flour are inducible by treatment with succinic anhydride. Gel electrophoretic and
solubility data show that there are alterations in the physical
and chemical properties of proteins, and in certain cases these
changes improve foam properties, that is, improve solubility and
polypeptide dissociation of proteins at the interface of the
foaming solution. Similar results have been reported for
succinylated soybean and sunflower seed proteins (44, 46).

Foam properties related to seed type. Soybean flour
suspensions produced thick egg white-type foams at all pH levels
tested except at 4.0 (Figure 14; 47). Although the increase in
capacity of suspensions at pH values 6.5 and 4.0 was identical,
a medium thick foam was produced by the latter. At pH 4.0, the
level of soluble protein in the suspension was significantly
lower than at the higher pH values; the latter three percentages
of protein were similar. A decline in foaming capacity at pH
8.2 was improved by the two-step adjustment from pH 6.7 to 4.0
to 8.2. Percentage of soluble protein in the suspensions did
not seem to be clearly related to the capacity of soybean
suspension to form foams.

Foam capacity of peanut seed flour suspension at pH 4.0,
6.7, and 8.2 was similar. The two-step pH adjustment to 8.2
produced a larger increase in capacity than the other
treatments. Foams produced by suspensions at pH 6.7 and 8.2
were similar, being of a thick egg white-type consistency.

Suspensions of field pea flour at pH 6.7 and 8.2 (including
the two-step adjustment) contained similar high quantities of
soluble protein; at pH 4.0, most of the protein was insoluble.
Foam capacity of suspensions was higher at pH 8.2 than at 4.0
and 6.7. The two-step pH adjustment did not improve foam
capacity over that of the one-step change as shown with the
soybean and peanut products. The foam produced at pH 4.0 was
thinner than those at pH 6.4 and 8.2; the latter three products
had similar consistencies.

Pecan flour suspensions exhibited poor foaming properties.
Not only were the capacities and soluble protein values low, but
the foams produced were thin and contained large, unstable air

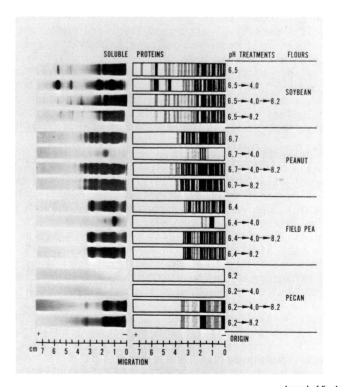

Journal of Food Science

Figure 15. Typical disc PAGE patterns of soluble proteins of defatted soybean, peanut, field pea, and pecan seed flour suspensions at various pH values (47)

cells. The suspension at pH 1.5 had the lowest amount of soluble protein but produced the highest foam capacity. Adjusting the pH from 6.7 to 8.2 directly or by the two-step adjustment improved foam capacity; no differences were noted between the two treatments at pH 8.2.

Observations made in this study indicate that protein solubility was more closely related to the type of foam produced than to increase in capacity. For example, soybean and peanut seed flour suspensions contained higher levels of soluble protein than field pea and pecan suspensions and produced foams of much thicker consistency and smaller air cells.

Data from gel electrophoretic studies suggest that the major seed-storage proteins are important in foam capacity (Figure 15; 47). Gel patterns of soybean flour suspensions at pH 4.0 contain a number of major (dark-staining) proteins. Proteins having slow mobility are usually the large molecular weight storage globulins of seeds. As suspension pH is increased to 6.7 and 8.2, solubility of these globulins is increased. In all cases, suspensions adjusted by the two-step procedure had similar patterns to those adjusted directly. Thus, differences in functionality of these two suspensions cannot be related to protein quality as distinguished by the gel electrophoretic techniques used in this study. However, these data suggest that solubility of the major storage proteins, or their subunit components, contribute to foam capacity. In addition, other seed constituents, such as carbohydrate and ash (for example, field pea and pecan, respectively) may be equally involved (especially when the suspension pH is 1.5).

Conclusions

Proteins have surfactant properties that play an important role in the functional behavior of food ingredients in foams. The proteins lower interfacial tension between gas and the aqueous phase, facilitating deformation of the liquid into an impervious film. That is, the film has enough viscosity and mechanical strength to prevent rupture and maintain coalescence. Basically, the proteins undergo conformational change at the interface with unfolding to expose hydrophobic regions that facilitate the association of the polypeptides while resisting migration into the water phase. These changes allow formation of a continuous cohesive film around the air vacuole. The level or proportion of certain water-soluble proteins is a measure of the availability of these components for functional expression in foams. Varying levels of protein, salt, pH, heat, and chemical denaturants, in addition to molecular interactions among various constituents in the ingredients, affect certain physicochemical properties of the proteins in suspension media used in foam formation. Proteins denatured to various dissociated or aggregated forms are detectable by changes in size, conformation, and charge, by polyacrylamide disc gel electrophoretic techniques.

Acknowledgments

The authors appreciate the assistance of C. H. Vinnett and J. I. Wadsworth on the research of the foam properties and multiple regression analyses, respectively, with glandless cottonseed flour. Names of companies or commerical products are given solely for the purpose of providing information; their mention does not imply recommendation or endorsement by the U.S. Department of Agriculture over others not mentioned.

Literature Cited

1. Ramsden, W. Trans. Liv. Biol. Soc., 1919, 33, 3.
2. Freundlich, H. "Colloid and Capillary Chemistry"; Methuen and Co., Ltd.: London, 1926; 884 p.
3. Claytôn, W. "Theory of Emulsions and Their Technical Teatment": J. and A. Churchill, Ltd.: London, 1928; p. 75.
4. Bikerman, J. J. "Foams: Theory and Industrial Applications" Reinhold Pub. Corp.: New York, 1953; p. 158.
5. Peter, P. N.; Bell, R. W. Ind. Eng. Chem., 1930, 22, 1124.
6. Cumper, C. W. N.; Alexander, A. E. Trans. Faraday Soc., 1950, 46, 235.
7. Cumper, C. W. N. Trans. Faraday Soc., 1953, 49, 1360.
8. Kitchener, J. A.; Cooper, C. F. Q. Rev., 1959, 13, 71.
9. Graham, D. E.; Phillips, M. C. In "Foams"; Akers, R. J., Ed.; Academic Press: New York, 1976; p. 237.
10. Mita, T.; Nikai, K.; Hiraoka, T.; Matsue, S.; Matsumoto, H. J. Colloid Interface Sci., 1977, 59, 172.
11. Horuichi, T.; Fukushima, D.; Sugimoto, H.; Hattori, T. Food Chem., 1979, 3, 35.
12. Mita, T.; Ishida, E.; Matsumoto, H. J. Colloid Interface Sci., 1978, 64, 143.
13. Kinsella, J. E. CRC Crit. Rev. Food Sci. Nutr., 1976, 7, 219.
14. Kinsella, J. E. J. Am. Oil Chem. Soc., 1979, 56, 242.
15. Eldridge, A. C.; Hall, P. K.; Wolf, W. J. Food Technol., 1963, 17, 20.
16. Yasumatsu, K; Sawada, K.; Moritaka, S.; Misaki, M.; Toda, J.; Wada, T.; Ishii, K. Agric. Bio. Chem., 1972, 36, 719.
17. McKelier, D. M. B.; Stadelman, W. J. Poult. Sci., 1955, 34, 455.
18. Waniska, R. D.; Kinsella, J. E. J. Food Sci., 1979, 44, 1398.

19. Lawhon, J. T.; Cater, C. M. J. Food Sci., 1971, 36, 372.
20. Buckingham, J. H. J. Sci. Food Agric., 1970, 21, 441.
21. Mangon, J. L. N.Z. J. Agric. Res., 1958, 1, 140.
22. Institute of American Poultry Industries. "Chemical and Bacteriological Methods of American Poultry Industries"; Chicago, Ill., 1956.
23. McWatters, K. H.; Cherry, J. P. J. Food Sci., 1975, 40, 1205.
24. Cherry, J. P.; Berardi, L. C.; Zarins, Z. M.; Wadsworth, J. I.; Vinnett, C. H. In "Nutritional Improvement of Food and Feed Proteins"; Friedman, M., Ed., Plenum Publ. Corp.: New York, 1978; p. 767.
25. Cherry, J. P.; McWatters, K. H.; Beuchat, L. R. In "Functionality and Protein Structure"; Pour-El, A., Ed.; ACS Symp. Series: Washington, D.C., 1979; p. 1.
26. Hermansson, A. M. J. Texture Stud., 1975, 5, 425.
27. Watts, B. M. Ind. Eng. Chem., 1937, 29, 1009.
28. Smith, A. K.; Schubert, E.; Belter, P. A. J. Am. Oil Chem. Soc., 1955, 32, 274.
29. Cherry, J. P.; McWatters, K. H. J. Food Sci., 1975, 40, 1257.
30. Cherry, J. P.; McWatters, K. H.; Holmes, M. R. J. Food Sci., 1975, 40, 1199.
31. McWatters, K. H.; Cherry, J. P.; Holmes, M. R. J. Agric. Food Chem., 1976, 24, 517.
32. Steel, R. G. D.; Torrie, J. H. "Principles and Procedures of Statistics"; McGraw - Hill Book Co., Inc.: New York, 1960; p. 227.
33. Holmes, M. R. In "Functional Properties Governing Roles of Proteins in Foods"; Cherry, J. P., Ed.; ACS Symposium Series: Washington, D.C., 1980.
34. DeVilbiss, E. D.; Holsinger, V. M.; Posati, L.; Pallansch, M. Food Technol., 1974, 28, 40.
35. Beuchat, L. R. J. Agric. Food Chem., 1977, 25, 258.
36. Gounaris, A. D.; Perlmann, G. J. Biol. Chem., 1967, 242, 2739.
37. Meighen, E. A.; Schachman, H. K. Biochemistry, 1970, 9, 1163.
38. Klotz, I. M.; Keresztes-Nagy, S. Biochemistry, 1963, 2, 445.
39. Chen, L. F.; Richardson, T.; Amundson, C. H. J. Milk Food Technol., 1975, 38, 89.
40. Gandhi, S. K.; Schultz, J. R.; Boughey, F. W.; Forsyth, R. H. J. Food Sci., 1968, 33, 163.
41. Groninger, H. S., Jr. J. Agric. Food Chem., 1973, 21, 978.
42. Groninger, H. S., Jr.; Miller, R. M. J. Food Sci., 1975, 40, 327.
43. Grant, D. R. Cereal Chem., 1973, 50, 417.
44. Franzen, K. L.; Kinsella, J. E. J. Agric. Food Chem., 1976, 24, 788.

45. Childs, E. A.; Park, K. K. J. Food Sci., 1976, 41, 713.
46. Canella, M.; Castriotta, G.; Bernardi, A. Lebensm. -
     Wiss. Technol., 1979, 12, 95.
47. McWatters, K. H.; Cherry, J. P. J. Food Sci., 1977, 42,
     1444.

RECEIVED September 5, 1980.

Water and Fat Absorption

C. W. HUTTON

Department of Food, Nutrition, and Institution Management,
The University of Alabama, University, AL 35486

A. M. CAMPBELL

Agricultural Experiment Station and College of Home Economics,
The University of Tennessee, Knoxville, TN 37916

This chapter does not purport to be the final word on water and fat absorption of plant proteins. Rather, it is designed to summarize information on the mechanism of protein interaction with water and fat, to pull together the various terms used to describe and methods used to assess water and fat absorption, and to encourage a more uniform and quantitative approach to the study of protein functionality and performance in food. The major protein products to be examined in this review are of soy origin; other products will be reviewed briefly in comparisons with soy products.

Protein is utilized in many foods for the particular characteristics that it contributes to the final product (1). In order for protein products to maintain or enhance the quality and acceptability of a food, the protein ingredients should possess certain functional properties that are compatible with the other ingredients and environmental conditions of the food system (2). Consequently, an important aspect of the development of new protein additives and their incorporation into food systems is the establishment of their functional properties. Functional properties of proteins are physicochemical properties through which they contribute to the characteristics of food. Study of functionality should provide information as to how a protein additive will perform in a food system (2, 3, 4). These properties are affected by protein source, composition, and structure; prior treatment; and interaction with the physical and chemical environment. It is generally believed that proteins are the principal functional component of plant additives. However, in products such as flours and meals, the carbohydrates also may play an active role in the functional performance of the additives.

Water absorption or hydration is considered by some as the first and the critical step in imparting desired functional properties to proteins. Most additives are in dehydrated form; the interaction with water is important to properties such as hydration, swelling, solubility, viscosity, and gelation. Protein has been reported to be primarily responsible for water absorption,

0097–6156/81/0147–0177$06.00/0

although other constituents of the additive have an effect. Primary protein-water interactions occur at polar amino acid sites on the protein molecules. Water retention of proteins is related to the polar groups, such as carbonyl, hydroxyl, amino, carboxyl, and sulfhydryl groups; most proteins contain numerous polar side chains along their peptide backbone, making them hydrophilic. Water binding varies with the number and type of polar groups (5). Other factors that affect the mechanism of protein-water interactions include protein conformation and environmental factors that affect protein polarity and/or conformation. Conformational changes in the protein molecules can affect the nature and availability of the hydration sites. Transition from globular to random coil conformation may expose previously buried amino acid side chains, thereby making them available to interact with aqueous medium. Consequently, an unfolded conformation may permit the protein to bind more water than was possible in the globular form (6).

Fat absorption of protein additives has been studied less extensively than water absorption and consequently the available data are meager. Although the mechanism of fat absorption has not been explained, fat absorption is attributed mainly to the physical entrapment of oil (7). Factors affecting the protein-lipid interaction include protein conformation, protein-protein interactions, and the spatial arrangement of the lipid phase resulting from the lipid-lipid interaction. Non-covalent bonds, such as hydrophobic, electrostatic, and hydrogen, are the forces involved in protein-lipid interactions; no single molecular force can be attributed to protein-lipid interactions (8).

Experimental Procedures

Water Absorption

Various terms and methods have been proposed in the literature to describe and evaluate the uptake of water by a protein ingredient. Unfortunately, few standardized methods exist for the evaluation of this functional property; most methods have developed piecemeal and are empirical. To describe the uptake of water, water absorption, water binding, water hydration capacity, water holding, swelling, and possibly other terms have been used. Frequently, the terminology is related to the method employed (9). These terms often are misleading and confusing in the interpretation of results. Consequently, some of the published data are of limited use. Currently, the methods for evaluating the uptake of water appear to have been narrowed to four as follows:

1) Relative humidity method. Water absorption is defined as the water absorbed by a dried protein powder with equilibration against water vapor at a known relative humidity. This method, also known as the equilibrium moisture content (EMC) method, was described first by Mellon et al. (10). Huffman et al. (11) used

the EMC method for measuring moisture absorption and retention of
five varieties of sunflower meal. As seen in Figure 1, little
difference was noted in water absorption over a wide range of
relative humidity (RH); the approximate was only 10-17% at RH up
to 70%. At higher RH, EMC rose rapidly to more than 30% at 90%
RH. Varietal differences were reported to affect EMC, and this
was attributed to differences in carbchydrate content. Yet, protein content was highly correlated with water absorption.

Hagenmaier (12) reported that proteins may be ranked in
order of water binding capacity at one relative humidity and that
this order should hold true at other relative humidities. He
evaluated water binding of several oilseed proteins at 84% RH;
water binding increased as the number of hydrophilic groups of
the different proteins increased.

2) Swelling method. Measurement of swelling is a second
method used to estimate water absorption of a sample. A swelling
apparatus was devised and described by Hermansson (2, 13). In
this method, a small amount of sample is dusted on a wet filter
paper fastened on a glass filter. The filter is fitted on top of
a thermostated funnel filled with water and connected to a horizontally located capillary. The uptake of fluid is followed in
the capillary; evaporative losses are prevented by a glass lid
(2). Swelling is defined as the spontaneous uptake of water by
the additive.

Most materials of interest as ingredients are neither completely soluble nor completely insoluble, and most foods are
water swollen systems. The concept of uptake of water or swelling may, therefore, provide valuable information for evaluation
of a protein as a food ingredient. Swelling was defined by
Hermansson (2) as

. . . the spontaneous uptake of a solvent by a
solid. It is a phenomenon frequently observed as
the first step in the solvation of polymers, in
which case swelling continues until the molecules
are randomized within the system. In other cases
solvation may be prevented by various intermolecular
forces in the swollen sample, resulting in limited
swelling and a definite volume increase.

Hermansson and co-workers (2, 13, 14) examined the swelling
ability of a soybean protein isolate (P-D), sodium caseinate,
whey protein concentrate (WPC) and fish protein concentrate (FPC)
in pure water and as influenced by pH and ionic strength. The
swelling characteristics of the various proteins were entirely
different. The differences included magnitude of swelling, the
time to reach maximum swelling, the tendency to solubilize, and
the response to the chemical environment. Generally, P-D exhibited very good swelling, which was considered to be a limited
type of swelling. Caseinate also showed a high degree of

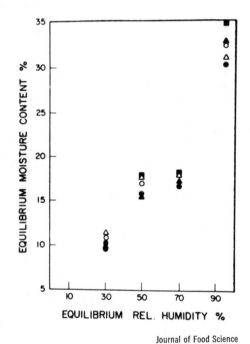

Figure 1.   Water absorption of sunflower meals.   Sunflower varieties: (△) Arrowhead; (●) Mingren; (▲) Greystripe; (○) Peredovik; (■) Krasnodarets (11).

swelling, which was of the unlimited type; that is, swelling was high prior to solvation. The WPC and FPC showed poor swelling ability. Hermansson (2) contends that swelling, together with solubility data, could be a useful guide to functionality.

A modification of the swelling apparatus was used to evaluate WPC, peanut protein, single cell protein (SCP), and chicken preen gland protein. Chicken protein exhibited the highest swelling values followed in decreasing order by peanut, SCP, and WPC products (15).

Problems associated with use of this method involve production of the apparatus and lack of reproducibility of the test conditions, specifically, application of samples in reproducible thickness to the wet filter paper (16).

3) <u>Excess water method</u>. A third method, which reportedly measures binding, involves exposure of the sample to excess water and application of mild force (usually centrifugal) to separate the retained or "bound" water from the free water. The amount of water retained, usually determined as the weight gain, is reported as water absorption expressed as a percentage of the dry sample weight. This method, to date, has been used most frequently by researchers (17, 18, 19, 20). The details of the specific procedures vary, as do the units of expressing water absorption. Other drawbacks are that this method, unless considerably modified, does not account for the protein and carbohydrate solubilized by the procedure or for the low-density components that float on the supernatant surface. For these reasons, some researchers (16) suggest that the excess water method is suitable only for use on protein additives that are mainly insoluble.

For measuring water absorption by the excess water method, the techniques developed by Janicki and Walczak (described by Hamm, 21) for meats and by Sosulski (22) for wheat flour are modified. Lin et al. (17) modified the Sosulski technique for use with sunflower and soy meal products. This modified procedure has been employed for much of the research on water absorption of plant protein additives. Water absorption capacities of a soy flour, two soy concentrates, and two soy isolates were compared by Lin et al. (17) to those of a sunflower flour, three sunflower concentrates, and one sunflower isolate. The percent water absorption of the soy products increased as the total protein content of the samples increased from flour to isolate. The soy flour absorbed 130% water, the soy concentrates absorbed an average of 212% water, and the soy isolates absorbed an average of 432% water. No calculations were made, however, that related the percent water absorbed to protein content of the samples. The sunflower products, though similar in protein content, did not respond in the same magnitude or direction as the soy products. All sunflower products had lower water absorption values than did their soy counterparts, and water absorption did not parallel protein content. The water absorption value for the isolate (155%) was within the range (138-203%) reported for the sunflower

concentrates. On the basis of these data, the researchers suggested that the soy proteins are more hydrophilic than the sunflower proteins.

Fleming et al. (18), using the procedure of Janicki and
Walczak, also compared the water absorption characteristics of
sunflower and soybean products. Water absorption of the soy
products increased as protein content of the samples increased.
Consistent with the findings of Lin et al. (17), water absorption
of the sunflower products was below that of the soy products
except for the concentrates, which had values similar to those of
the soy counterpart, and the water absorptions of the sunflower
concentrates and isolates were of similar magnitude.

The water absorption of alfalfa leaf protein (ALP) extracted
and prepared by different methods was compared to that of a soy
concentrate and a soy isolate (19). The method for determining
water absorption was that of Lin et al. (17) and Fleming et al.
(18); units for water absorption were g H_2O/g leaf protein. The
values reported for the ALP provide information concerning the
effect of processing on water absorption. Of more interest to
this reviewer, however, is the researchers' use, as standards for
water absorption, of two soy products, a concentrate, Promosoy-
100, and an isolate, Promine-D, which have been evaluated by
other researchers employing the excess water/centrifuge method.
Hutton (23) and Hutton and Campbell (20) examined the effect of
pH and temperature on the water absorption of these same two soy
products using a modification of this method. Values for water
absorption for these products as compiled from the various studies
are recorded in Table I. The same products were used, and a similar (similar only in terms of referenced method being excess
water) procedure was used. The conditions of the test undoubtedly
were modified for the particular laboratories or researchers or to
be compatible with other measurements or variables evaluated in
the respective studies. In summary, within the conditions of a
particular test, all researchers reported, with one exception,
a greater water absorption by the isolate than the concentrate,
possibly indicating a relationship between water absorption and
protein content. This has been suggested by several investigators. The one exception, as reported by Hutton (23), was that
the water absorption of the concentrate was greater than that of
the isolate at pH 5.0. At pH 5.0, the isoelectric point (IEP) of
the protein is being approached. Perhaps, at the IEP, the protein has less of an effect and the hydrophilic carbohydrates have
proportionately more of an effect on water absorption than at
other pH levels. Possibly the relationship between water absorption and solubility, yet to be discussed, accounts in part for the
statement by Quinn and Paton (16) that the excess water method is
suitable for products that are mainly insoluble.

Hutton (23) and Hutton and Campbell (20) used the excess
water method in combination with the solubility measurements but
did not adjust the absorption data for the solubilized protein,

which was quite different for the two products. A recalculation of these data to adjust the absorption values for the solubilized protein resulted in the comparisons of original and adjusted values shown in Tables II and III for a soy protein concentrate and isolate, respectively. Solubilized carbohydrate was not considered in the adjusted values.

Taking the loss of soluble protein into account in calculating water absorption can influence the conclusions drawn in some studies. Elgedaily and Campbell (24) used the excess water method (23) in studying the effects of sodium chloride and sucrose on the water absorption of three soy protein isolates. They also determined nitrogen solubility and used the values for an alternative method of calculation in which the absorption values reflected a correction for the extracted protein. With both methods of calculation it was assumed that added sodium chloride and sucrose were extracted in the determination of soluble nitrogen. With the method of calculation in which the extracted protein was disregarded, the analysis of variance showed soy isolate, level of sodium chloride, and their interaction to significantly affect water absorption. When the correction was made, only the soy isolate appeared to have a significant effect on water absorption.

In a current study in the same laboratory (25), the correction is being made by determining the total solids in the supernatant. This approach might be preferable in a study in which other ingredients are added to the soy products.

4) Water saturation method. More recently, a method has been described by Quinn and Paton (16), who claim that their procedure more closely simulates actual food production applications than does the excess water method. In this technique, only enough water, essentially all of which is retained upon centrifugation, is added to saturate the sample. A comparison of the excess water method and the proposed method for various protein products, as reported by Quinn and Paton, is presented in Table IV.

Values for water absorption by the Quinn and Paton method are different from values obtained in the same laboratory by the excess water method, particularly for the more soluble protein materials; however, water absorption values determined by the "same" method by various investigators are not always similar either (Table I).

Quinn and Paton (16) observed that the intensity and duration of mixing influenced the water uptake of the materials investigated, and that the measurement was not reproducible unless the time of agitation was constant (2 min was chosen). One product, sodium caseinate, required longer than a 2-min mixing time. These investigators state that the proposed technique measures water absorbed and retained under specific conditions, which may or may not apply to particular manufacturing applications. They do, however, state that this method with the application of limited

Table I. Water absorption values for a soy concentrate and isolate as reported in several studies.

Reference Number	Concentrate (Promosoy-100)	Isolate (Promine-D)
	----------------- % -----------------	
17	196	417
18	--	775[a,b]
18		415[a,c]
19	330[a]	638[a]
23	340[d]	1218[d]
23	273	350[e]
23	241	190[f]

[a]Values reported were converted to percent absorption.
[b]Sample mixed 1 min.
[c]Sample mixed 10 min.
[d]pH 7.0.
[e]pH 6.0.
[f]pH 5.0.

Table II. Water absorption mean values determined by excess water method and as adjusted for solubilized protein for a soy concentrate (23).

pH	Excess water method			Adjusted for solubilized protein		
	5.0	6.0	7.0	5.0	6.0	7.0
	------------------- % ----------------------					
Temp (°C)						
4	247	268	326	251	278	340
ambient[a]	241	273	340	248	281	358
90	305	306	537	320	342	733

[a]22-25°C.

Table III. Water absorption mean values determined by excess water method and as adjusted for solubilized protein for a soy isolate (23).

	Excess water method			Adjusted for solubilized protein		
pH	5.0	6.0	7.0	5.0	6.0	7.0
	------------------- % ----------------------					
Temp (°C)						
4	175	457	975	183	492	1845
ambient[a]	190	350	1218	204	475	2500
90	408	493	1158	440	703	4958

[a]22-25°C.

Table IV. Water hydration capacity values of various protein materials (16).

	Excess water method[a]	Proposed method[b]
Pea concentrate	1.05	1.31
Promosoy-100 concentrate	3.10	3.00
Promine-D isolate	3.50	3.85
Supro 620 isolate	6.70	5.50
Rapeseed concentrate	4.50	3.29
Caseinate	0.00	2.33
Egg white	1.30	0.67
Whey concentrate	0.00	0.97

[a]Method of Fleming et al. (18).
[b]Values were obtained before the technique was standardized. The values are estimates of where the WHC lies within the experimentally determined range.

rather than excess water more closely simulates actual food
product conditions. Certain drawbacks of the method are evident.
Water absorption has been shown by Quinn and Paton and by others
to be affected by time of exposure of product to water; the
controlled limited exposure time in this method may make the
values more reproducible, but does it simulate actual performance
conditions? Also, this method eliminates the problem of solubilized protein; the solubilized protein may have an important
effect on water absorption by being no longer available to be
hydrated.

Fat Absorption

For fat absorption, the measurements have been fewer and
more consistent than those reported for water absorption. Therefore, one should be able to examine with more security the data
of separate studies. The main problem associated with the fat
absorption method appears to be that rarely, if ever, is a food
system encountered that involves only the protein-lipid
interaction.

Fat absorption of protein additives usually is measured by
adding excess liquid oil to a protein powder, mixing and holding,
centrifuging, and determining the amount of absorbed oil (total
minus free) (17, 26). The amount of oil and sample, kind of oil,
holding and centrifuging conditions, and units of expression have
varied slightly from one investigator to another; consequently,
fat absorption values for the same products have varied among
investigators. Values for Promosoy-100 and Promine-D, as
reported in three separate studies, which were most consistent in
methodology, are reported in Table V. Variations do exist in the
various studies to be reviewed, fat absorption of proteins is
affected by protein source, extent of processing and/or composition of additive, particle size, and temperature.

Water Absorption Response Patterns

Considering the limitations presented thus far that are
inherent in interpreting data reported in the literature, perhaps
the pattern of response rather than the individual measurements
is of importance in predicting functionality. Information will
be presented that relates water absorption to other functional
properties and examines the effect of the physical and chemical
environment on the water absorption response patterns of various
protein ingredients. This presentation will be brief and with
limited explanation. Original references may be consulted by
those desiring greater depth.

Relationships with other properties. In any food system and
possibly in simple systems, the protein ingredient is likely to
perform several functions, most of which are being discussed in

detail individually in this book. Since water binding or absorption may be a result of other properties, it seems of interest to mention some relationships. Hermansson (27) states that "water-binding may be caused by any of the following properties: (a) the ability to swell and take up water; (b) a high viscosity caused by soluble molecules, swelled particles or a mixture; (c) the ability to form a gel network during processing." Inherent in this explanation of water binding are the associations of the absolute water absorption (previously discussed), solubility, viscosity, and gelation functions. In Table VI, data are presented that demonstrate the effect of temperature on the solubility, swelling, and viscosity properties of a soy isolate.

Solubility: The reported relationship between water absorption and solubility of proteins has not been consistent. Water absorption capacity of sunflower concentrates increased slightly as the solubility index of the protein decreased (17). Hermansson (2) reported that a highly soluble protein exhibits poor water binding, but a reverse relationship between water absorption, evidenced by swelling and solubility, was not observed. In a later report, Hermansson (27) stated that solubility measurements give no information as to whether or not a protein will bind water.

Hagenmaier (10) demonstrated that pH had little effect on water absorption of oilseed protein products, but solubility was pH dependent. He suggested that the differing degree of dependence on pH indicates that water absorption and protein solubility are not correlated. Contrastingly, Wolf and Cowan (28) reported the pH-water retention curve of soy proteins to follow the pH-solubility curve. Both solubility and water retention were minimal at the isoelectric point (4.5) and increased as the pH diverged from this point. Hutton and Campbell (20) reported that the effects of pH and temperature on water absorption of soy products paralleled those of solubility for the most part. At pH 7.0, as temperature increased from 4 to 90°C, solubility increased but water absorption increased then decreased. This suggested that water absorption and solubility may be related to a point, perhaps maximum hydration, at which solubility continues to increase and hydration does not. This appears consistent with the statements of Hermansson (2) that water absorption is the first step in the solvation of polymers, and swelling may be limited or unlimited.

Viscosity and gelation: Many proteins absorb water and swell, causing changes that are reflected by concurrent increases in viscosity (9). Viscosity has been reported to be influenced by solubility and swelling (2, 19). As water absorption (determined as swelling) increased, viscosity also increased (2). Fleming et al. (18) reported that water absorption was attributable to the protein content of the product and that viscosity increased exponentially as protein content increased, thereby suggesting a possible relationship between

Table V.   Fat absorption values for a soy concentrate
and isolate as reported in several studies.

Reference Number	Concentrate	Isolate
17	92	119
19	101[a]	156[a]
23	74	121

[a]Reported values were converted from ml oil/g to % by
multiplying by specific gravity of oil and multiplying
by 100.

Table VI.   Effect of temperature on some functional properties
of soy isolate[a] (7).

Temperature (°C)	Solubility (%)	Swelling (ml/g)	Viscosity 15s^{-1}
25	53	10	
70	67	17	3620
80	68	20	7490
90	71	17	5280
100	81	14	1410

[a]Measurements were made at 25°C after heat treatment.

water absorption and viscosity. Hutton (23) reported that above pH 5.0, the water absorption of soy isolates was greater than that of soy concentrates, and that viscosity was greater for the isolate than the concentrate. However, pH and temperature variations brought about different responses in water absorption and viscosity between the concentrate and isolate. This suggests that water absorption and viscosity are not always correlated, at least in products of quite different proximate composition.

A gel can be considered as a structural matrix holding liquid and can be formed spontaneously by swelling at high protein concentrations (7, 27). Yet no studies have been reviewed that related water absorption and gelation. The gelation phenomenon of soybean proteins has been studied in detail by Catsimpoolas and Meyer (29, 30, 31).

Environment. The physical and chemical environments have been shown to affect the functional performance of proteins. Factors, such as concentration, pH, temperature, ionic strength, and presence of other components, affect the balance between the forces underlying protein-protein and protein-solvent interactions (9). Most functional properties are determined by the balance between these forces. Although the comparison of discrete data from various studies might be of limited value, consideration of the response patterns of protein additives to changes in the environment of simple and/or food systems might be fruitful.

For products of dissimilar composition, e.g., isolates, concentrates, and flours, changes in the chemical environment undoubtedly elicit different responses from the various constituents of an ingredient, i.e., protein, carbohydrate, etc. Perhaps the properties of the protein can best be examined in the isolate because of its relatively low concentration of nonprotein constituents. Unfortunately, isolates usually have been subjected to the most extensive processing, which also affects the response of the protein.

Concentration: Generally, as product concentration increases, water absorption increases. Much controversy exists concerning what actually is being measured in water-unlimited and water-limited systems. Obviously, as product concentration increases to the point that water is limited, water absorption ceases to increase.

Hansen (32) reported that as protein content increased from 32 to 90%, water absorption of the soy samples increased. This can be seen in Figure 2. Although the trend is evident, the increase was not linear, which suggests that factors in addition to protein were contributing to water absorption. (Particle size did not affect absorption.)

As protein content increased from soy flour to soy isolate, water absorption also increased. This was not the case for sunflower products (17).

pH: Structurally, ionized amino acid groups bind more water than nonionized groups. Variation in pH affect the ionization of amino acid groups. Since amino acid composition and arrangement vary with protein origin, the response to pH also varies (6). Hagenmaier (12) reported that pH changes from 4.5 to 7.0 had little effect on cottonseed, soybean, and casein protein samples. The effect, though slight, was increased water binding as pH increased; the differences among the samples were greater than the change with pH. Similarly, Hermansson (14) demonstrated that swelling of a soy isolate, caseinate, and WPC increased as pH increased from 4.0 to 10.0. The increase was more dramatic for caseinate than either the isolate or WPC. Wolf and Cowan (28) reported that water absorption of soy proteins was minimal at the isoelectric point (4.5) and increased as the pH diverged from this point.

The effects of pH on water absorption of a soy isolate and soy concentrate are represented in Table VII. Water absorption of a soy isolate increased as pH increased from 5.0 to 7.0. This increase was evident at all temperatures studied (4°C, ambient, and 90°C). For soy concentrate samples, the effect of pH was similar in direction to the effect on water absorption of the isolate but was of smaller magnitude (23).

Fleming et al. (18) did not examine the effect of pH on water absorption, but these researchers examined the effect of "pH activation" on water absorption of sunflower and soy products. For pH activation, 1.25 N NaOH was added to slurries to achieve pH 12.2 and then 6.0 N HCl was added to return to pH 6.0 in 10 min. The pH activation process improved the water absorption properties for most products but did not increase water absorption of the soy flour. Processes similar to pH activation may be encountered in the processing of vegetable protein additives.

Temperature: Proteins usually bind less water at high temperatures than at low temperatures, but if protein conformation changes with heating, it could override the effect of temperature on water absorption. Heating, concentrating, drying, and texturizing may all cause denaturation and aggregation of protein molecules, which may reduce the surface area and availability of polar amino acid groups for water binding. Protein aggregation, in some cases, may change the conformation in such a way as to increase binding. All protein additives are subjected to denaturation to some extent; therefore, response of products to temperature will be affected by product origin and prior processing treatment (6).

Hermansson reported that heating of a soy isolate enhanced swelling. Huffman et al. (11) reported that temperature had no noticeable effect on water absorption of sunflower products.

Hutton and Campbell (20) reported that the overall effect of temperature, disregarding pH, was increased water absorption for a soy isolate and a trend in that direction for a concentrate. The effect of temperature on water absorption of the soy isolate

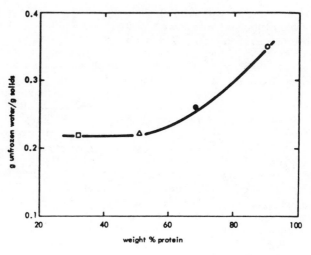

Figure 2. Water binding (grams of unfrozen water/gram of solids) by soy protein preparations, each containing 1 g of water/g of solids, as a function of protein content, where % protein (on a solids basis) = % N × 6.25: (○) soy protein isolate B; (●) soy protein concentrate; (△) soy flour (defatted); (□) carbohydrate-enriched fraction of soy concentrate (32).

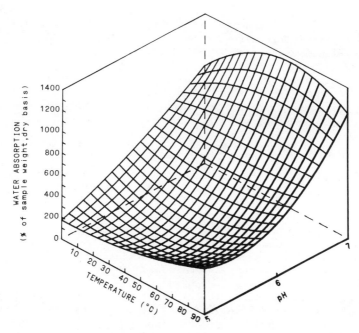

Figure 3. Water absorption response surface for Promine-D with variations in pH and temperature (20)

was less than that of pH and was pH dependent. This can best be
viewed in the response surface, Figure 3. At pH 5.0, water
absorption increased as temperature increased. At pH 6.0, water
absorption decreased, then increased with increasing temperature.
At pH 7.0, water absorption increased then decreased with increasing temperature. The response surface for the concentrate
is shown in Figure 4 (20). For the soy concentrate, the effect
of temperature was less than that of pH, and water absorption
increased as temperature increased from 4 to 90°C.

Ionic strength: Salts compete with water for the
binding sites on amino acid side groups; the amount of water
bound to a protein is a function of salt concentration (6). The
effect of ionic strength on protein functionality has been
focused primarily on its effect on solubility (2). Generally,
protein solubility increases at low salt concentrations and
decreases at high salt concentrations.

Hermansson (2) examined the effect of ionic strength on
swelling of soy isolate, caseinate, and WPC products. The effect
of NaCl on water absorption was not consistent among the products.
Generally, as ionic strength increased from 0.0 to 1.0 M NaCl,
water absorption increased for the soy isolate and caseinate.
Water absorption of WPC was little affected by ionic strength.
Fleming et al. (18) examined the effect of 5% NaCl on the
water absorption of soy and sunflower flours, concentrates, and
isolates. The water absorption of the soy and sunflower flours
was higher in 5% NaCl than in water. Generally, salt decreased
water absorption of the isolates; the concentrates of both plant
products were little affected by NaCl. Data reported by Fleming
et al. reflect a response to NaCl; however, the type of ion is
known to affect the type of response of other properties (33)
and possibly the same is true for water absorption.

Sucrose: Elgedaily and Campbell (24) found sucrose,
within the concentration range used, to have no significant
effect on water absorption of three soy protein isolates.

Fat Absorption Response Patterns

The values for the soys were presented first (Table V) since
most investigators used soy products as a reference for comparison of the fat absorption of other plant products. Lin et al.
(17) examined the fat absorption capacities of various sunflower and soy products. All sunflower products (flour, concentrates, and isolates) bound more oil than the soy counterpart.
Fat absorption for the sunflower products ranged from 130 to
448% of their weight on a 14% moisture basis; fat absorption
values for the soys ranged from 84 to 154%. The sunflower proteins appear to be more structurally lipophilic than the soy
proteins. It seemed likely to the investigators that the sunflower proteins contain numerous nonpolar side chains that are
believed to bind hydrocarbon chains, thereby contributing to

increased oil absorption. (Personal observation of the average reported data for the flour, concentrate, and isolate categories within sunflower and soy products revealed that fat absorption increased as protein content increased, sunflower products exhibiting greater fat absorption. Also within the group of soy isolates and concentrates studied, the percent fat absorption decreased as water solubility of the product increased. These possible relationships were not discussed.)

Wang and Kinsella (19) studied the fat absorption, and other properties of alfalfa leaf protein (ALP) concentrate and used the soy protein concentrate and isolate Promosoy-100 and Promine-D, respectively, as the references. The fat absorption values are reported in ml oil/g sample. Converting these to percent fat absorbed (based on the specific gravity of peanut oil) results in values that are higher than those reported by Lin et al. (17); this was most evident in the case of the isolate.

All ALP products absorbed more oil than the soy products. Acetone treatment of ALP resulted in reduced fat absorption; ALP with higher lipid contents absorbed more oil. ALP extracted with water and NaOH absorbed more oil than those extracted with NaCl or Tris buffer. These researchers (19) attribute fat absorption to physical entrapment; a correlation of 0.95 was found between fat absorption and bulk density. However, more oil was absorbed by the ALP than the soy products even though the products had similar bulk densities.

Fat absorption of rapeseed products was compared to that of soy flours and concentrates (34). All rapeseed products exhibited fat absorption values greater than those of soy products. For both rapeseed and soy, fat absorption of concentrates was greater than that of the flour and meals. Values for the rapeseed isolate were similar to those for the rapeseed concentrate.

Textured soy flours were reported to have oil absorption values that ranged from 65 to 130% of their dry weight with small particles absorbing more oil than large ones. The maximum fat absorption occurred within 20 min for all particles.

Hutton (23) and Hutton and Campbell (26) evaluated the fat absorption capacities of a soy concentrate and isolate as a function of temperature. (Values obtained for the ambient temperature treatment, 22-25°C, are reported in Table V.) Fat absorption was expressed on per gram of sample, as-is moisture, dry-weight, and per gram of protein bases as shown in Table VIII. Fat absorption of the soy isolate (P-D) was greater than that of the concentrate (P-100) for all bases of expression, but expression in terms of protein content brought the fat absorption values of the two products closer together. This suggests that the fat absorption was attributable primarily to the protein. It also suggests that the additional carbohydrate did not absorb as much oil as the protein and that the absorption by the protein was different for the two products. (That is, the protein was

Table VII. Percent water absorption values of a soy concentrate
            and isolate as a function of ph (23).

	Concentrate[a]	Isolate
pH	----------------- % -----------------	
5.0	270	257
6.0	282	433
7.0	402	1117

[a]Weight of concentrate equal to weight of isolate.

Table VIII. Fat absorption of a soy concentrate and isolate at
             22-25°C (23).

Expression basis	Fat absorption		
	P-D	P-100₁[a]	P-100₂[b]
Sample wt., as-is	121.1	72.4	75.5
Sample wt., dry	129.8	77.5	80.8
Protein wt.	134.4	108.3	112.9

[a]Equal sample weight relative to P-D.
[b]Equal protein weight relative to P-D.

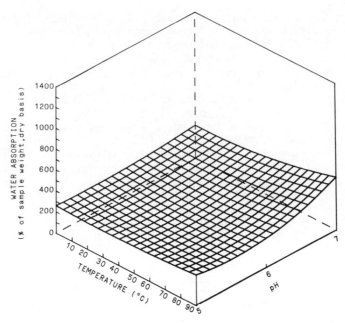

Figure 4.   Water absorption response surface for Promosoy-100$_2$ with variations in pH and temperature (20)

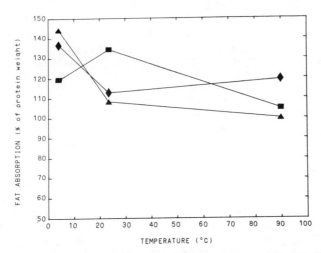

Figure 5.   Fat absorption (percent of protein weight) of Promosoy-100$_1$ (equal sample weight relative to Promine-D), Promosoy-100$_2$ (equal protein weight relative to Promine-D), and Promide-D held at 4°C, ambient temperature (22°–25°C) and 90°C: (▲) P-100$_1$; (♦) P-100$_2$, (■) P–D (26).

primarily responsible, but either the presence of carbohydrate and/
or other constituents in the concentrate or differences in
processing of the isolate and concentrate resulted in a slightly
different response by the protein of the two products.)

The effect of temperature and the interaction of temperature
and soy were significant, as was soy product. The effect of
temperature on fat absorption is shown in Figure 5. Overall,
as temperature decreased from 90 to 4°C, fat absorption
increased. This could be attributable to increased viscosity of
the oil at decreased temperatures. Possibly, the high viscosity
of the system contributed to greater ease of physical entrapment.
Or perhaps the higher temperatures denatured the protein in some
way that resulted in decreased fat absorption. There was a
temperature-soy interaction; the isolate and concentrate did not
respond similarly to temperature variations. The isolate exhibited maximum fat absorption at ambient temperature; the concentrate exhibited a maximum at 4°C. This suggests that some-
thing other than protein is contributing to fat absorption.

Food Product Performance

The performance of protein additives in food products is
the ultimate test of functionality (7). Functional performance,
particularly moisture absorption, has been used as a criterion
for selection of protein additives for food systems. Yet this
has for the most part been a trial-and-error situation. The
majority of data on water and fat absorption of plant protein
additives in food systems has involved the incorporation of such
products into comminuted meat systems. In relation to water and
fat absorption, meat systems are an excellent test system.
Several researchers (35, 36, 37, 38) report that when soy products are combined with ground meat products, total (moisture and
fat) cooking losses are reduced as percent substitution of the
additive is increased. Seideman et al. (39), on the other hand,
reported that beef patties with added soy retained more moisture
but lost more fat during cooking than did all-beef patties.

In products like sausage and meat pies, plant protein
additives have been shown to improve binding of the structure and
reduce moisture and fat losses. This may be related to the
gelling and/or emulsification functions, as well as, water and
fat absorption.

Protein additives also find use in numerous baked products.
Soy flour increases the shelf life and freshness of breads, pancakes, waffles, and cakes through enhanced moisture retention
(28, 40). Addition of soy flour to pancakes and doughnuts helps
prevent excessive fat absorption during cooking (28). This
phenomenon of decreased fat absorption is not well understood.

As presented, data are available on assessment of water
absorption in simple systems and on the incorporation of protein
additives into some food systems. Functional properties in

simple systems as related to performance in food systems have been studied to a lesser degree (2, 3, 15, 26, 41, 42).

The most detailed studies were reported by Hermansson and Åkesson (3, 41) and Hermansson (42) in which the properties of a soy isolate, caseinate, WPC, and model test systems of additive and lean beef or pork were studied. Solubility, swelling, and viscosity (properties reviewed as related to water absorption) were correlated with moisture loss in the raw systems. In cooked systems, the best predictability of meat texture as affected by additive was a statistical model that included the functional properties of swelling and gel strength of protein additive dispersions.

Torgersen and Toledo (15) assessed the relationship between solubility and water absorption of WPC, peanut protein, single cell protein (SCP), and chicken preen gland protein and characteristics in a comminuted meat system containing the various protein additives. Protein additives with high water absorption capacities were correlated with more viscous raw mixtures and with less fat and water released on cooking. Simple system determination of water absorption capacity at 90°C rather than at 78.5°C was the better predictor of water and fat retention in the food system studied.

Conclusions

The wide array of plant protein products, methods, and terms used to evaluate and describe functional properties makes it difficult to compare products and results of separate studies. In high protein products, water and fat absorption is attributable primarily to the protein. In other products, additional ingredient components have an effect. Hydrophilic polysaccharides greatly affect water absorption. Water absorption also is affected by product concentration, pH, temperature, and ionic strength. Fat absorption has been studied less extensively than water absorption but has been reported to be affected by temperature, size of ingredient particles, and degree of denaturation of the protein. For both properties, functional performance of the protein product may be modified greatly by ingredient and/or food system constituents. Plant protein applications in food systems have included bakery and dairy products, salad dressings, snack foods, and ground meat. The effects of plant protein products in meat systems are related most often to water and fat absorption functions. However, predictability of food system performance by simple system functionality tests is not inevitable.

It is apparent that the subject of water and fat absorption by plant proteins is characterized and complicated by interrelatedness. The study of relationships is important and is complicated by problems of measurement; comparisons between studies are hindered by the variety of methods and conditions

used in different laboratories. Some standardization should be
undertaken. A few studies in which sorption in simple systems
has been related to performance in food systems have been
reported; more are needed. The effects of other food ingredients
on functional properties of proteins should be further
investigated. Obviously much remains to be done.

Literature Cited

1. Hammonds, T. M.; Call, D. L. Agr. Econ. 320, Cornell
    University Agr. Exp. Sta., Ithaca, N. Y., 1970.

2. Hermansson, A. -M. "Proteins in Human Nutrition," Ch. 27,
    Porter, J. W. G., and Rolls, B. A., Eds., Academic
    Press, London, 1973.

3. Hermansson, A. -M.; Åkesson, C. J. Food Sci., 1975, 40,
    595.

4. Johnson, D. W. J. Am. Oil Chem. Soc., 1970, 47, 402.

5. Kuntz, I. D., J. Am. Chem. Soc., 1971, 93, 514.

6. Chou, D. H.; Morr, C. V. J. Am. Oil Chem. Soc., 1979, 56,
    53A.

7. Kinsella, J. E. J. Am. Oil Chem. Soc., 1979, 56, 242.

8. Ryan, D. S. Adv. Chem. Series, 1977, 160, 67.

9. Kinsella, J. E. Crit. Rev. Food Sci. and Nutr. 1976, 7,
    219.

10. Mellon, E. F.; Korn, A. H.; Hoover, S. J. Am. Oil Chem.
    Soc., 1947, 69, 827.

11. Huffman, V. L.; Lee, C. K.; Burns, E. E. J. Food Sci.,
    1975, 40, 70.

12. Hagenmaier, R. J. Food Sci., 1972, 37, 965.

13. Hermansson, A. -M. Lebensm.-Wiss. U. Technol., 1972, 5,
    24.

14. Hermansson, A. -M.; Sivik, B.; Skjöldebrand, C. Lebensm.
    -Wiss. U. Technol., 1971, 4, 201.

15. Torgersen, H.; Toledo, R. T. J. Food Sci., 1977, 42, 1615.

16. Quinn, J. R.; Paton, D. Cereal Chem., 1979, 56, 38.

17. Lin, M. J. Y.; Humbert, E. S.; Sosulski, F. W. J. Food
     Sci., 1974, 39, 368.

18. Fleming, S. E.; Sosulski, F. W.; Kilara, A.; Humbert, E. S.
     J. Food Sci., 1974, 39, 188.

19. Wang, J. C.; Kinsella, J. E. J. Food Sci., 1976, 41, 286.

20. Hutton, C. W.; Campbell, A. M. J. Food Sci., 1977, 42, 454.

21. Hamm, R. Adv. Food Res., 1960, 10, 355.

22. Sosulski, F. W. Cereal Chem., 1962, 39, 344.

23. Hutton, C. W. "Functional properties of a soy isolate
     and a soy concentrate in simple systems as related
     to performance in a food system"; Dissertation, The
     University of Tennessee, Knoxville, Tenn., 1975.

24. Elgedaily, A.; Campbell, A. M. Unpublished data.

25. James, C.; Campbell, A. M. Unpublished data.

26. Hutton, C. W.; Campbell, A. M. J. Food Sci., 1977, 42,
     457.

27. Hermansson, A. -M. J. Am. Oil Chem. Soc., 1979, 56, 272.

28. Wolf, W. J.; Cowan, J. C. Crit. Rev. Food Technol., 1971,
     2, 81.

29. Catsimpoolas, N.; Meyer, E. W. Cereal Chem., 1970, 47,
     559.

30. Catsimpoolas, N.; Meyer, E. W. Cereal Chem., 1971, 48,
     150.

31. Catsimpoolas, N.; Meyer, E. W. Cereal Chem., 1971, 48,
     159.

32. Hansen, J. R. J. Agric. Food Chem., 1978, 26, 301.

33. Anderson, R. L.; Wolf, W. J.; Glover, D. J. Agr. Food
     Chem., 1973, 21, 251.

34. Sosulski, F.; Humbert, E. S.; Bui, K.; Jones, J. D.
     J. Food Sci., 1976, 41, 1349.

35. Adolphson, L. C.; Horan, F. E. Cereal Sci. Today, 1974,
     19, 441.

36. Anderson, R. H.; Lind, K. D. Food Technol., 1975, 29,
        (2), 44.

37. Drake, S. R.; Hinnergardt, L. C.; Kluter, R. A.; and Prell,
        P. A. J. Food Sci., 1975, 40, 1065.

38. McWatters, K. H. J. Food Sci., 1977, 42, 1492.

39. Seideman, S. C.; Smith, G. C.; Carpenter, Z. L. J. Food
        Sci., 1977, 42, 197.

40. Levinson, A. A.; Lemancik, J. F. J. Am. Oil Chem. Soc.,
        1974, 51, 135A.

41. Hermansson, A. -M.; Åkesson, C. J. Food Sci., 1975, 40,
        603.

42. Hermansson, A. -M. J. Food Sci., 1975, 40, 611.

RECEIVED October 21, 1980.

Emulsifiers: Milk Proteins

C. V. MORR

Department of Food Science, Clemson University, Clemson, SC 29631

Emulsions, such as those in food products, may be defined as macroscopic dispersions of two immiscible liquids, one of which forms the continuous, dispersion phase and the other, the discontinuous, dispersed phase, commonly termed globules. An emulsion of two immiscible liquids, one polar and one nonpolar, will rapidly separate into two distinct phases upon standing unless a third phase, an adsorbed surfactant, is present in the interface to stabilize it. Surfactants, e.g., chemical emulsifiers and proteins, stabilize emulsions by protecting against close contact and the association of individual globules. The DLVO theory (1) ascribes emulsion stability to a balance of attractive van der Waals forces and electrostatic repulsive forces, derived from oppositely charged ions in a double layer surrounding the globules (Figure 1). The emulsion remains stable so long as the magnitude of the repulsive forces exceeds that of the attractive forces between the globules.

There are a number of factors that influence the stability of emulsions: 1) interfacial tension between the two phases; 2) characteristics of the adsorbed film in the interface; 3) magnitude of the electrical charge on the globules; 4) size and surface/volume ratio of the globules; 5) weight/volume ratio of dispersed and dispersion phases; and 6) viscosity of the dispersion phase. Four classes of emulsion stabilizing agents can be distinguished: 1) inorganic electrolytes; 2) surface active agents or surfactants; 3) finely divided insoluble solids; and 4) macromolecular emulsifying agents, such as proteins, gums and starches (2).

The immense interfacial area separating dispersed globules from the dispersion phase is of critical importance in determining their stability. For example, it is estimated that a typical emulsion has approximately 7×10^5 cm^2 interfacial area per liter (3). Thus, those factors controlling the properties of the interfacial membrane are extremely important in determining the stability of the emulsion.

Chemical and electron microscopic techniques have been used to verify the presence of an adsorbed layer surrounding emulsion globules. Proteins are examples of hydrocolloids that exhibit

0097–6156/81/0147–0201$05.00/0

unique surfactant properties due to their large molecular weights
and their multiplicity of hydrophobic and hydrophilic residues,
each of which exhibit a spectrum of affinities for the polar and
non-polar phases in the emulsion system. The polymer structure of
protein molecules adsorbed in the interfacial layer provide multiple attachment sites (4) and are believed to adsorb in a manner
that exposes amino acid segments to the aqueous phase (1). Each
protein system exerts its own unique influence on emulsion stability, since each possesses a characteristic complement of amino
acids, arranged in a specific sequence that controls its ability
to adsorb in the interface. For these reasons, proteins provide a
variety of responses to compositional factors, e.g., pH, ionic
composition, chemical emulsifiers, and processing treatments.
Proteins contribute to emulsion stability under those compositional
and processing conditions that favor their own stability, but also
produce emulsion instability under adverse compositional and processing conditions which reduce their stability.Under proper conditions proteins provide an indispensable stabilizing function in
emulsions which are subsequently subjected to a variety of food
processing treatments such as freeze-thaw, dehydration-rehydration,
ultra-high temperature (UHT) sterilization, and others.

Milk Proteins as Emulsifiers in Milk Systems

Properties of milk fat globules. Milk fat globules (MFG) range
in size from 0.1 to 10 μm diameter, with an average of 2 to 4 μm
and are surrounded by a protective membrane (MFGM), which consists
of phospholipids, lipoproteins, caseins and immunoglobulins (5,6).
Several models have been proposed for the structure of MFGM in
freshly drawn milk (5,6,7). King's model (7) depicts a highly
ordered structure with successive layers of high melting triglycer-
ides, phospholipids and adsorbed proteins. Harper and Hall (5)
present a somewhat more simplified model with an intermediate
phospholipid layer that provides polar regions for attachment of
strongly-bound, insoluble proteins surrounded by an outer layer of
loosely-bound, soluble proteins (Figure 2). The loosely-bound
protein layer is readily removed by treating with detergents and
by processing treatments that provide sufficient shear forces,
such as mixing and churning. The most tightly bound proteins in
the MFGM are highly associated, hydrophobic lipoproteins whose
origin is membrane material derived from secretory processes in
the mammary system (5). The proteins and phospholipids in the MFGM
contribute substantially to the stability of the natural MFG by
controlling the degree of clustering of the globules via an agglutination mechanism similar to that of bacterial cells, but does not
contribute significantly to emulsion stability in homogenized milk
products. Casein micelles, adsorbed from the aqueous phase, contribute most to the stability of derived MFG in homogenized milk
products (8), but phospholipids also contribute to their stability.
     Homogenization is an indispensable processing treatment for

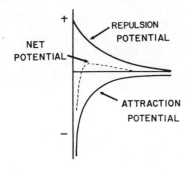

Figure 1. DLVO Theory for explaining emulsion stability

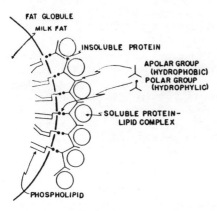

AVI Publishing Company, Inc.

Figure 2. Diagram of milk fat globule membrane (5)

milk products that require high emulsion stability, e.g., fluid homogenized milk, evaporated milk, ice cream and others. The small size of the MFG in these products (0.1 to 1.0 μm) is important in that it results in a large surface/volume ratio and enhanced protein adsorption, which protects against association of the MFG and further minimizes their tendency to cream by increasing their buoyant density. The uniformity of size also minimizes their tendency for varying rates of rise, as predicted by Stokes' Law (3).

The MFG-casein micelle complex that forms during homogenization is stabilized by van der Waals and hydrophobic bonding (6,10). Various workers have confirmed the association of casein micelles and their subunits at freshly formed MFG during homogenization (8). Berger (11) proposed that casein micelles are disrupted during high pressure treatments such as during homogenization. This interpretation is logical in view of the previous findings of Schmidt and Buchheim (12) that casein micelles are effectively disrupted under high pressure treatments.

Milk protein system. The nomenclature and physico-chemical properties of the major milk proteins and their subunits have been provided by Whitney et al. (13) and Brunner (14). The conformation and related properties of the individual proteins and their subunits and aggregates have been reviewed by Morr (15) with special reference to their functional properties in food systems, and drawing heavily upon previous considerations by Bloomfield and Mead (16) and Slatterly (17).

The caseins, which represent about 80 % of the total proteins in milk, and are precipitated from milk by adjusting the pH to their isoelectric points (4.5-5.0),are composed of three major components, α_s-, β- and κ-caseins, each of which has been further fractionated into a number of subfractions (13). The caseins are present in milk as an equilibrium between soluble, submicellar complexes and the larger, spherical aggregates, termed micelles. This equilibrium is dependent upon pH, ionic activities (mainly Ca and PO_4 ions), temperature and other factors (18). The caseins are associated with a fairly constant proportion of inorganic colloidal phosphate which contributes substantially to their structural arrangement and integrity. Casein micelles in milk range in size from about 100 to 250 nm diameter, exhibit sedimentation constants of 8 to 22 X 10^2 S, molecular weights of 2 to 18 X 10^8 Daltons, and are composed of from 450 to 10,000 casein subunits (15,17). A number of models have been proposed for these complex structures (15) and even though they remain the subject of intensive research by milk protein chemists (19), the model presented in Figure 3 accomodates most of the present knowledge of the micellar system.

Even though casein micelles are remarkably stable in milk under normal conditions, they are quite susceptible to minor alterations in pH, ionic composition and to treatment with the enzyme rennin which cleaves the glycomacropeptide (GMP) portion of the κ-

casein component of the micelle (18). Of special interest to the
present consideration is the observation that casein micelles are
effectively disrupted by application of high pressures (12) and it
may be postulated that such disruptions probably involve the rupture of hydrophobic bonding that is so important in stabilizing
the micellar structure.

Whey proteins, e.g., α-lactalbumin, β-lactoblobulin, bovine
serum albumin and the immunoglobulins, remain soluble when milk is
treated with acid or rennin to precipitate the caseins. The whey
proteins, other than the euglobulin component of the immunoglobulins, do not play a significant role in stabilizing the MFG in
milk. Euglobulin is reported to play an important function in the
clustering of MFG due to an agglutination reaction similar to that
for bacterial cells. The whey proteins exert a pronounced influence
on the physico-chemical properties of the casein micelles (18)by
their strong tendency to interact with the κ-casein subunits in the
micelle through disulfide interchange reactions when subjected to
processing conditions that promote protein denaturation. It may
be presumed that such protein-protein interactions probably alter
the ability of the casein micelles to be disrupted by high pressure
treatments and might thereby inhibit their ability to stabilize
MFG in certain milk products subjected to high heat processing.

Attempts to isolate and characterize MFGM proteins have not
been completely successful due to their strong tendency to associate in most protein dissociating buffers. Thus, their role in
stabilizing milk emulsions is not entirely understood (6,14), even
though it is assumed that they are an integral part of the MFGM
system.

Milk Proteins as Emulsifiers in Food Systems

Milk protein products. As indicated in Table 1, the food
industry is placing major emphasis on the production and utilization of milk protein products in a wide variety of formulated food
products (20,21,22). Although nonfat dry milk (NFDM) and whey
powder are major milk protein ingredients in formulated foods,
casein and whey protein concentrates, which contain their proteins
in a more highly concentrated and functional form, are essential
for certain food product applications, such as those products that
require the proteins as an emulsifier agent. Additional details on
the processing methods and conditions used to produce the various
milk protein products are available (23).

NFDM, which retains casein micelles similar to those in fresh
milk, is produced by pasteurization of skimmilk, vacuum concentration and spray drying under processing conditions that result in
either "low heat" or "high heat" product. Low heat NFDM is required
for most applications that depend upon a highly soluble protein,
as the case for most emulsification applications, since it is
manufactured under mild temperature conditions to minimize whey
protein denaturation and complexation with casein micelles.

Table 1. U.S. Production/Utilization of
Milk Protein Products in Human Foods - 1978[a]

Product	Amount, million #	Protein Content, %, dry basis
Non-Fat Dry Milk	928.8	36
Casein & Caseinate	75-90	95
Concentrated Whey Solids	118.2	13
Dry Whey	534.7	13
Partly Delactosed Whey	32.2	20
Partly Demineralized Whey	28.6	13+
Whey Protein Concentrate	8.9	50
Whey Solids in wet blends	34.4	13

[a] Statistical data provided by:American Dry Milk Institute,
Chicago, IL; Whey Products Institute, Chicago, IL; and New
Zealand Milk Products, Inc., Rosemont, IL.

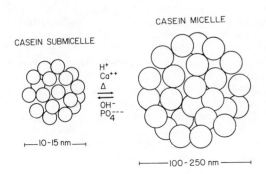

Figure 3.   Proposed model for casein micelles and submicelles

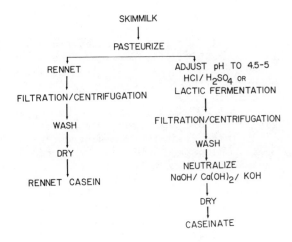

Figure 4.   Production scheme for casein and caseinate

Casein isolates, produced as caseins and caseinates, are obtained by two procedures: rennet treatment and acid precipitation (Figure 4). Rennet casein is produced by treating skimmilk with the enzyme rennin to cleave the GMP component of the κ-casein subunit and thereby render the entire casein micelle system unstable. The product represents an enzymatically modified protein product, which otherwise closely resembles the native casein micelle system in milk. It contains the original Ca content of the native micelles, but is insoluble and nonfunctional as an emulsifier unless the Ca ion content is diminished. Acid casein is produced by treating skimmilk with hydrochloric or sulfuric acids, or by fermentation with a lactic acid culture to lower the pH to isoelectric point conditions to precipitate the caseins. Casein micelles are completely dissipated under the latter treatment and the casein subunits undergo a high degree of protein-protein interaction to form precipitated particles that are easily recovered by filtration or centrifugation. Sodium and potassium caseinates, and calcium caseinates are produced by neutralization with the appropriate alkali to pH's of 8-10 to solubilize the casein precipitate, and then spray dried. Both sodium and potassium caseinates are totally soluble in water and form viscous solutions that exhibit excellent functionality in a wide variety of food products which require emulsification. On the other hand, calcium caseinate is virtually insoluble and performs rather poorly in most emulsifier applications (21).

Partly delactosed whey is produced by concentrating cheese whey or casein whey sufficiently to exceed the solubility limit of lactose, followed by cooling, seeding with lactose crystals and removal of the crystalline lactose. The resulting liquor fraction is recovered and dried.

Partly demineralized whey is produced by subjecting whey to electrodialysis or ion exchange processing treatments to prederentially remove polyvalent ions, vacuum concentrated and spray dried.

Whey protein concentrates (WPC) are produced by a variety of processing treatments to remove both lactose and minerals (20) as indicated in Figure 5. Even though it would be highly desireable to remove most of the lactose and minerals in these processes, it is not practical from an economic standpoint and thus most of these products only range in protein content from 35 to 50 %. The major objective of most of these processes is to produce a WPC with minimal protein denaturation in order to obtain a product with maximum protein solubility and functionality. However, from a practical consideration this objective is not readily obtainable, and thus most WPC products commercially available exhibit variable whey protein denaturation and functionality (20).

Lactalbumin is an insoluble whey protein product produced by heating whey to high temperatures (> 90 C) to denature and render the proteins insoluble when adjusted to isoelectric conditions by the addition of acid. These proteins offer little functionality in emulsification applications.

Co-precipitate is an insoluble milk protein product that is produced by heating skimmilk to high temperatures (> 90 C) to denature the whey proteins and complex them with the casein micelles. The heated system is subsequently adjusted to isoelectric point conditions of pH 4.5-5 to precipitate the complexed whey protein-casein micelles, centrifuged or filtered to recover the precipitate, washed and dried. The resulting product, which is virtually insoluble, exhibits only minor functionality in most typical emulsification applications.

Emulsification properties. Caseins and caseinates are commonly selected for food product applications that require surfactant properties, e.g., emulsification and foam stabilization, since they contain high protein contents of > 90 %, are highly soluble, and are resistant to heat-induced denaturation in products to be subjected to high temperature processing conditions (15).

Consideration of the conformational states of the major casein subunits (Figure 6) reveals that they consist of a non-uniform distribution of polar and hydrophobic residues along their polypeptide chains (16,17) leading to an amphiphilic molecule (17). These amphiphilic casein subunits exhibit similar primary structures among the different caseins, e.g., α_S-, β-, and κ-, that account for their excellent surfactant properties (15). These casein subunits are highly susceptible to interactions vial hydrophobic and ionic bonding, and thus exist in the form of highly associated aggregates that require highly alkaline pH conditions, removal of Ca ions, and the use of strong protein dissociating against such as urea and others to obtain complete monomerization (24). It should therefore be assumed that caseins and caseinates exist and function in most food applications in their highly associated form which possess the ability to release subunits as needed to stabilize emulsion globules as they are formed in homogenization processing (15). If it is assumed that those factors that favor dissociation of casein complexes favor its functionality as an emulsifier, it may be possible to predict what compositional and processing conditions would optimize functionality. However, such adjustments in processing conditions would also affect the intrinsic .properties of the emulsion system. Thus, it becomes obligatory, in most instances, to investigate and determine experimentally the specific conditions that should be selected to optimize the functionality of caseins in food emulsion systems.

Emulsification properties in model food systems. Pearson et al. (25) investigated the emulsification properties of caseinate and NFDM in model emulsion systems produced by blending soybean oil into an aqueous buffer system as a function of pH and ionic strength (Figures 7 and 8). They found that caseinate exhibited good emulsification properties under all pH and ionic strength conditions studied, but was particularly effective at pH 10.4.

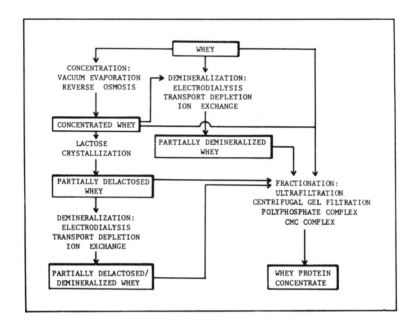

Food Technology

Figure 5.   Process flowchart for preparing whey protein concentrate (20)

---

10   20
H.Arg- Pro- Lys–His – Pro - Ile – Lys– His- Gln -Gly–Leu-Pro -Gln-‾Glu– Val – Leu –Asn–Glu –Asn–Leu-‾
                                                               Absent in variant A

30   40
‾Leu- Arg–Phe–Phe –Val –Ala‾–Pro– Phe- Pro –Gln–Val– Phe –Gly –Lys–Glu–Lys–Val –Asn–Glu–Leu-

50   60
Ser –Lys –Asp -Ile – Gly- Ser – Glu- Ser – Thr– Glu–Asp–Gln –Ala‾– Met–Glu –Asp-Ile – Lys –Glu– Met-
                                Þ   Þ   ThrP (variant D)

70   80
Glu –Ala –Glu–Ser –Ile – Ser – Ser – Ser –Glu–Glu–Ile – Val –Pro–Asn–Ser –Val –Gln–Lys –His –
           Þ   Þ   Þ   Þ   Þ

90   100
Ile – Gln –Lys –Glu–Asp –Val –Pro– Ser – Glu–Arg –Tyr –Leu–Gly –Tyr– Leu–Glu–Gln– Leu–Leu–Arg–

110   120
Leu–Lys –Lys –Tyr –Lys –Val –Pro –Gln–Leu–Glu–Ile – Val –Pro –Asn –Ser – Ala –Glu–Glu–Arg –Leu–
                                                                 Þ

130   140
His - Ser – Met- Lys- Gln–Gly–Ile – His – Ala –Gln– Gln– Lys –Glu–Pro –Met–Ile – Gly –Val –Asn–Gln–

150   160
Glu - Leu –Ala –Tyr –Phe –Tyr– Pro –Glu–Leu–Phe – Arg –Gln–Phe –Tyr– Gln– Leu–Asp-Ala –Tyr–Pro–

170   180
Ser– Gly –Ala –Trp –Tyr- Tyr –Val - Pro– Leu–Gly –Thr– Gln–Tyr –Thr–Asp - Ala –Pro– Ser – Phe –Ser–

190   199
Asp -Ile – Pro- Asn–Pro –Ile – Gly–Ser –Glu–Asn–Ser –‾Glu‾– Lys –Thr –Thr–Met–Pro– Leu–Trp.OH
                                                       Gly (variant C)

American Chemical Society

Figure 6.   Primary structure of bovine casein α_s-CN-B (15)

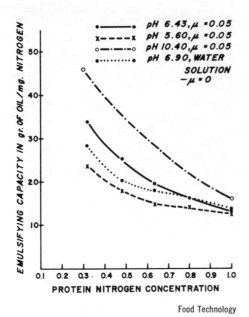

Food Technology

*Figure 7.   Emulsification properties of
potassium caseinate (25)*

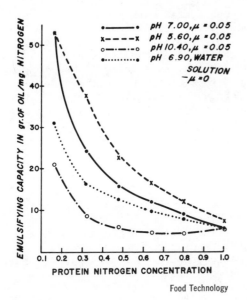

Food Technology

*Figure 8.   Emulsification properties of
NFDM (25)*

The emulsification properties of NFDM were slightly better than
for caseinate at all protein levels. However, NFDM exhibited lowest emulsification properties at pH 10.4 and highest emulsification at pH 5.6, which was directly opposite the results with
caseinate. Thus, the molecular state of caseins, whether in the
micellar or soluble complex form is important in determining their
functionality as an emulsifier.

Subharwal and Vakaleris (26,27) investigated the emulsion
stabilizing ability of caseinate in model emulsion systems containing added chemical emulsifiers to provide a range of HLB values
(Figures 9 and 10). The emulsification properties of caseinate
were concentration dependent, exhibiting maximum emulsification
at 0.2 to 0.4 % caseinate in the absence of chemical emulsifiers,
but exhibited emulsification impairment at 0.5 % caseinate in the
presence of HLB 11 emulsifiers. Addition of Ca and HLB 3-5 emulsifiers improved emulsion stability, whereas, addition of citrate
reduced emulsion stability. These findings are consistent with the
concept that Ca and citrate ions influence emulsion stability by
altering the relative extent of caseinate dissociation and in
addition they undoubtedly influence emulsion stability by their
effects on the charge and other inherent properties of the emulsion
globules, per se.

Whey protein concentrates (WPC), which are relatively new
forms of milk protein products available for emulsification uses,
have also been studied (4,28,29). WPC products prepared by gel
filtration, ultrafiltration, metaphosphate precipitation and
carboxymethyl cellulose precipitation all exhibited inferior emulsification properties compared to caseinate, both in model systems
and in a simulated whipped topping formulation (28). However,
additional work is proceeding on this topic and it is expected
that WPC will be found to be capable of providing reasonable
functionality in the emulsification area, especially if proper
processing conditions are followed to minimize protein denaturation during their production. Such adverse effects on the functionality of WPC are undoubtedly due to their irreversible interaction
during heating processes which impair their ability to dissociate
and unfold at the emulsion interface in order to function as an
emulsifier (22).

Conclusions

The following factors appear to control the emulsification
properties of milk proteins in food product applications: 1) the
physico-chemical state of the proteins as influenced by pH, Ca
and other polyvalent ions, denaturation, aggregation, enzyme
modification, and conditions used to produce the emulsion; 2) composition and processing conditions with respect to lipid-protein
ratio, chemical emulsifiers, physical state of the fat phase,
ionic activities, pH, and viscosity of the dispersion phase surrounding the fat globules; and 3) the sequence and process for
incorporating the respective components of the emulsion and for
forming the emulsion.

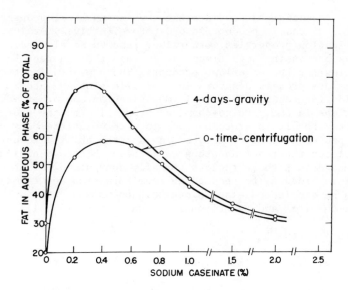

Journal of Dairy Science

Figure 9.   Emulsification properties of sodium caseinate in the absence of chemical emulsifiers (26)

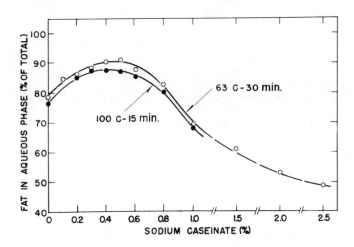

Journal of Dairy Science

Figure 10.   Emulsification properties of sodium caseinate in the presence of HLB 11 emulsifiers (26)

It is essential to consider the physico-chemical properties of
each WPC and casein product in order to effectively evaluate their
emulsification properties. Otherwise, results merely indicate the
previous processing conditions rather than the inherent functional
properties for these various products. Those processing treatments
that promote protein denaturation, protein-protein interaction via
disulfide interchange, enzymatic modification and other basic
alterations in the physico-chemical properties of the proteins
will often result in protein products with unsatisfactory emulsification properties, since they would lack the ability to unfold at
the emulsion interface and thus would be unable to function. It is
recommended that those factors normally considered for production
of protein products to be used in foam formation and foam stabilization be considered also, since both phenomena possess similar
physico-chemical and functionality requirements (30,31).

Literature Cited

1. Friberg, S., "Food Emulsions", Marcel Dekker, Inc., New York,
   N.Y. 1976.
2. Cante, C.J.; Franzen, R.W.; Saleeb, F.Z. J. Am Oil Chem. Soc.,
   1979, 56, 71A.
3. Jenness, R.; Patton, S."Principles of Dairy Chemistry", Wiley
   and Sons, New York, N.Y., 1959.
4. Tornberg, E.; Hermannson, A.M. J. Food Sci., 1977, 42, 468.
5. Harper, W.J. In "Dairy Technology and Engineering" W.J. Harper
   and C.W. Hall, Eds. AVI Pub. Co., Inc., Westport, Conn., 1976.
6. Brunner, J.R. In "Fundamentals of Dairy Chemistry", B.H. Webb
   and A.H. Johnson, Eds. AVI Pub. Co., Inc., Westport, Conn.,
   1965.
7. King, N. "The Milk Fat Globule Membrane", Comm. Agr. Bur.,
   Farnham Royal, Bucks, England, 1955.
8. Graf, E.; Bauer, H. In "Food Emulsions", S. Friberg, Ed., Marcel
   Dekker, Inc., New York, N.Y., 1976.
9. Ogden, L.V.; Walstra, P.; Morris, H.A. J. Dairy Sci., 1976, 59,
   1727.
10. Fox, K.K.; Holsinger, V.; Caha, J.; Pallansch, M.J. J. Dairy
    Sci., 1960, 43, 1396.
11. Berger, K.G. In "Food Emulsions", S. Friberg, Ed. Marcel
    Dekker, Inc., New York, N.Y., 1976.
12. Schmidt, D.G.; Buchheim, W. Milchwissenschaft, 1970, 25, 596.
13. Whitney, R. McL.; Brunner, J.R.; Ebner, K.E.; Farrell, H.M.,
    Jr.; Josephson, R.V.; Morr, C.V.; Swaisgood, H.E. J. Dairy Sci.,
    1976, 59, 795.
14. Brunner, J.R. In "Food Proteins", J.R. Whitaker and S.R. Tannen-
    baum, Eds. AVI Pub. Co., Inc., Westport, Conn., 1977.
15. Morr, C.V. In "Functionality and Protein Structure", A. Pour-El,
    Ed. American Chemical Society, Washington, D.C., 1979.

16. Bloomfield, V.A.; Mead, R.J., Jr. J. Dairy Sci., 1975, 58, 592.
17. Slatterly, C.W. J. Dairy Sci., 1976, 59, 1547.
18. Morr, C.V. J. Dairy Sci., 1976, 58, 977.
19. Payens, T.A.J. J. Dairy Res., 1979, 46, 291.
20. Morr, C.V. Food Technol., 1976, 30, 18.
21. Morr, C.V. J. Dairy Res., 1979, 46, 369.
22. Morr, C.V. New Zealand J. Dairy Sci. and Technol., 1979, 14, 185.
23. Fox, K.K. In "Byproducts from Milk", B.H. Webb and E.O. Whittier, Eds. AVI Pub. Co., Inc., Westport, Conn., 1970.
24. von Hippel, P.H.; Waugh, D.F. J. Am. Chem. Soc., 1955, 77, 4311.
25. Pearson, A.M.; Spooner, M.E.; Hegarty, G.R.; Bratzler, L.J. Food Technol., 1965, 19, 1841.
26. Subharwal, K.; Vakaleris, D.G. J. Dairy Sci., 1972, 55, 277.
27. Subharwal, K.; Vakaleris, D.G. J. Dairy Sci., 1972, 55, 283.
28. Morr, C.V.; Swenson, P.E.; Richter, R.L. J. Food Sci., 1973, 32, 324.
29. Wit, J.N. de; Boer, R. de Zuivelzicht, 1976, 68, 442.
30. Cooney, C.M. Dissertations Abstracts International, 1975, B36, 1123.
31. Richert, S.H.; Morr, C.V.; Cooney, C.M. J. Food Sci., 1974, 39, 142.

RECEIVED September 5, 1980.

11

Emulsification: Vegetable Proteins

KAY H. McWATTERS

Department of Food Science, University of Georgia College of Agriculture
Experiment Station, Experiment, GA 30212

JOHN P. CHERRY

Southern Regional Research Center, Science and Education Administration,
Agricultural Research, U.S. Department of Agriculture,
P.O. Box 19687, New Orleans, LA 70179

Emulsifiers are classified as a group of surface-active agents that can stabilize a dispersion of two immiscible liquids such as water and oil. In an emulsion, the dispersed droplets are commonly referred to as the disperse, discontinuous, or internal phase, and the medium in which they are suspended is the continuous, or external phase. The most common types of emulsions are oil-in-water (e.g., mayonnaise or milk) and water-in-oil (e.g., butter or margarine) and are classified according to the component that comprises the disperse phase. The terms "oil" and "water" are used in a general manner; almost any highly polar, hydrophilic liquid falls into the "water" category whereas hydrophobic, nonpolar liquids are considered "oils" (1). In the realm of food emulsions, the classic definition has been broadened to include two-phase systems of dispersions of gas in a liquid (e.g., foams) and combinations of liquids, solids, and gases in emulsions or batters (2).

Theories of Emulsions and Emulsifier Mechanisms

Several detailed discussions have described the complex theories of emulsion technology (1, 3, 4, 5). To summarize these theories, emulsifiers are essential for emulsion formation and stabilization to occur; these surface-active compounds reduce the surface and interfacial tensions between two immiscible liquids, but this property accounts for only part of the mechanisms at work in emulsification. Three separate mechanisms that appear to be involved in formation of a stable emulsion include: 1) reduction of interfacial tension, 2) formation of a rigid interfacial film, and 3) electrical charges.

Molecules in liquids exhibit different behaviors, depending upon their location in the volume of liquid. Those in the interior are surrounded by other molecules having balanced attractive forces. Those at or near the surface, however, are only partially surrounded by the other molecules of the liquid, and the attractive forces operating on them are unbalanced. There is a net

0097–6156/81/0147–0217$06.50/0

attraction toward the main body of the liquid as the liquid tends
to reduce its surface to a minimum, thus exhibiting the behavior
of the surface known as surface tension. Surface-active agents
are important in this regard because they are positively adsorbed
at the surface of immiscible components and reduce the surface
tension between them. The manner in which the molecules of
surface-active agents arrange themselves depends upon the distribution of electrons forming the bonds between the atoms of the
molecule or portion of the molecule. Hydrophilic (water-loving)
molecules have asymmetrically placed electrons and exhibit polar
properties; hydrophobic or lipophilic molecules have symmetrical
arrangements and exhibit nonpolar properties. Surface-active
agents are compounds that contain both hydrophilic and hydrophobic
groups and therefore exhibit polar-nonpolar character. In effect,
a monolayer that prevents close contact between droplets of the
disperse phase is created; a surface-active agent aids in preventing droplet coalescence.

In addition to lowering surface tension, surface-active
agents contribute to emulsion stability by oriented adsorption
at the interface and by formation of a protective film around the
droplets. Apparently, the first molecules of a surfactant introduced into a two-phase system act to form a monolayer; additional
surfactant molecules tend to associate with each other, forming
micelles, which stabilize the system by hydrophilic-lipophilic
arrangements. This behavior has been depicted by Stutz et al. (6)
and is shown in Figures 1-5.

Petrowski (7) summarized the effects of electrical charge
or balance between attractive and repulsive forces of particles
on emulsion stability. The repulsive forces, electrostatic in
nature, are stabilizing because they tend to keep droplets
separated, thereby preventing coalescence. The arrangement of
the electrical charges at the interface has been the subject of
some controversy; it has been described as a double layer of
opposite-charged ions or as a single, immobile counter-ion layer
surrounded by a diffuse layer of mobile ions extending outward.
The repulsive interaction of droplets possessing like electrical
charge seems to contribute to emulsion stability. Electrical
charges on droplets in emulsions can arise by ionization, absorption, or frictional electricity produced by the large shearing
forces required for emulsion formation.

Viscosity is an important physical property of emulsions in
terms of emulsion formation and stability (1, 4). Lissant (1) has
described several stages of geometrical droplet rearrangement and
viscosity changes as emulsions form. As the amount of internal
phase introduced into an emulsion system increases, the more
closely crowded the droplets become. This crowding of droplets
reduces their motion and tendency to settle while imparting a
"creamed" appearance to the system. The apparent viscosity continues to increase, and non-Newtonian behavior becomes more marked.
Emulsions of high internal-phase ratio are actually in a "super-
creamed" state.

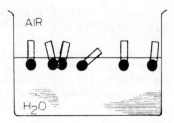

Food Product Development

Figure 1. Orientation of surfactant molecules at water–air interface (6)

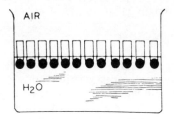

Food Product Development

Figure 2. Monolayer coverage (6)

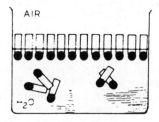

Food Product Development

Figure 3. Beginning of micelle formation (6)

Figure 4. Micelle cross section (6)

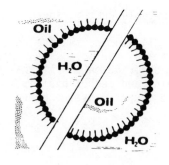

Food Product Development

Figure 5. Water-in-oil emulsion stabilized by a surfactant (upper left) and oil-in-water emulsion stabilized by a surfactant (lower right) (6)

Proteins constitute an important group of emulsifiers because they behave in a manner similar to that of surface-active agents by forming strong monolayer films at the interface. Their structure, and hence their behavior, is more affected by such variables as salt concentration, pH, and temperature than would occur with the use of conventional emulsifiers (7). Vegetable proteins, along with those derived from dairy and meat sources, have been intensively studied in recent years in efforts to assess their potential as functional ingredients. Because several studies and reviews describe emulsifying properties of soy protein in detail (8-15), the emphasis in this discussion will be directed toward vegetable proteins other than soy.

Experimental Procedures

The technique most commonly employed for measuring emulsion capacity of proteins is the model system, or modifications of it, developed by Swift et al. (16). This procedure involves the continuous addition of oil or melted fat to a protein dispersion during high-speed mixing; oil is added until a sudden drop in emulsion viscosity occurs due to separation of oil and water into two phases. The volume of oil added until the "breakpoint" is reached is used to express emulsion capacity of a protein; these values may be expressed as total ml of oil emulsified, ml of oil per unit weight of sample, or ml of oil per unit of protein or nitrogen in the sample. Comparisons of results obtained from different studies is difficult because small variations in technique, equipment, blender speeds, rate of oil addition, protein source and concentration, temperature, or type of oil affect emulsifying properties of proteins (17, 18, 19). Model systems are useful tools, however, because conditions and ingredients can be well defined and carefully controlled; they are also simpler and more economical to use than more complicated systems. No standardized tests exist for evaluating the emulsifying properties of proteins, and in many cases there seems to be little correlation between results obtained in model systems and those obtained in performance trials in food systems. The importance of designing model test systems that simulate actual food-processing conditions as closely as possible has been emphasized by Puski (20) for obtaining meaningful information regarding emulsifying properties of proteins.

Emulsifying Properties of Selected Vegetable Protein Sources

Peanut Seed. Ramanatham et al. (21) studied the influence of such variables as protein concentration, particle size, speed of mixing, pH, and presence of sodium chloride on emulsification properties of peanut flour (50% protein) and peanut protein isolate (90% protein). Emulsions were prepared by the blender

method previously described (16); distilled water at pH 6.8
comprised the continuous phase. For those tests in which flour
concentration was held constant, as in the particle size and
mixing speed tests, a level of 5 g flour dispersed in 50 ml distilled water at pH 6.8 was used.

Data in Table I show that emulsion capacity of peanut flour
decreased with increasing flour or protein concentration while
emulsion viscosity increased. This phenomenon was also demonstrated by McWatters and Holmes (22). A decrease in flour particle
size increased emulsion capacity and viscosity appreciably. Increasing the rate of mixing, however, decreased emulsion capacity
but increased viscosity. Increased speeds produce greater shear
rate, which decreases the size of the oil droplet; thus, there is
an increase in the surface area of the oil to be emulsified by
the same amount of soluble protein (23, 24).

Data in Figure 6 show the effect of varying the pH and sodium
chloride concentration on emulsion capacity of peanut protein
isolate. Shifting the pH to levels above or below the isoelectric
point improved emulsion capacity of peanut protein isolate in 0.1M
or 0.2M NaCl. Similar trends were noted when distilled water was
used as the continuous phase (data not shown). At the 0.5M NaCl
concentration, however, little difference was noted in emulsion
capacity at pH 3, 4, or 5; appreciable increases occurred when the
pH was raised to 6 and above. At the highest salt concentration
(1.0M NaCl), a gradual increase in emulsion capacity occurred when
the pH was increased from 3 to 10. An overall suppression in
emulsion capacity occurred as salt concentration increased except
at pH 5 and 6. These emulsion-capacity curves closely resemble
the protein-solubility curves of peanut protein shown in Figure 7
(25), indicating that factors typically influencing peanut protein
solubility also influence emulsion capacity. Thus, altering the
electrovalent properties by shifts in pH as well as changing the
ionic environment of peanut protein modifies its solubility and
emulsion capacity (21, 22, 26, 27). Whereas the salt-soluble
proteins seem to be the most functional in terms of emulsion
capacity of meat proteins (17), those that are soluble in water
or weak salt solutions are the most functional in this respect
where peanut proteins are concerned.

Attention has been directed toward modifying functional
properties of peanut proteins by chemical, enzymatic, and
physical approaches. Chemical modification has included
acetylation and succinylation treatments (28, 29). Marked improvement in emulsion capacity occurred as a result of this treatment if the proteins were extracted in acid (28). Beuchat et al.
(30) and Sekul et al. (31) found that enzymatic hydrolysis adversely affected emulsification properties of peanut protein.
However, another study by Beuchat (32) utilized a higher concentration of peanut flour than reported earlier (30) and showed
that pepsin hydrolysis at pH 2.0 (22° and 50°C) resulted in
emulsion capacities exceeding those of the untreated control;

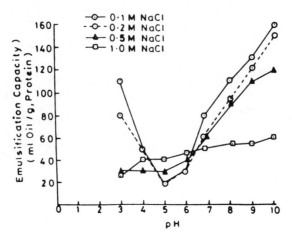

Figure 6.   Effect of sodium chloride at different concentrations on emulsification
              capacity of peanut protein isolate at various pHs (21)

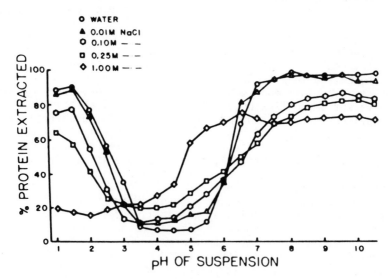

Figure 7.   Effect of sodium chloride at different concentrations on the extraction of
              peanut protein at various pHs (25)

emulsions formed from pepsin and bromelain hydrolyzates were low viscosity, whereas trypsin hydrolyzate emulsions were thick and viscous. Fungal fermentation of peanut flour reportedly decreased emulsion capacity (33), but the effects of using higher concentrations of peanut flour or continuous phases other than salt solution are not known. The application of moist heat has been employed with certain vegetable proteins as a physical approach to improving nutritional and flavor qualities; consequently, functional properties also undergo modification during heating. Emulsion capacity of whole full-fat peanuts was not adversely affected by heating the seeds in water (34), although significant changes in protein solubility and structural components occurred at high temperatures (35). Steam heating peanuts in the form of a finely ground flour, however, reduced levels of soluble nitrogen slightly but sharply reduced emulsion capacity (36). Thus, processing conditions and methods can have profound effects on protein character and functionality.

Pea and Bean Seeds. McWatters and Cherry (27) investigated the influence of pH adjustment on emulsion capacity and viscosity of cowpea flour (24.2% protein, dry wt basis) using the procedure of Carpenter and Saffle (23). Data in Table II show that adjusting the pH from the natural level of 6.4 to 4.0 reduced emulsion capacity by about 20%; adjusting the pH from the natural level to 8.2 resulted in a 15% increase in emulsion capacity. A two-step pH adjustment from 6.4→4.0→8.2 produced little increase (4%) in emulsion capacity over the unadjusted sample and was not as effective as the one-step pH adjustment to 8.2 in improving emulsion capacity. This flour at the natural pH level produced emulsions that were similar in consistency to a commercial mayonnaise. Flour suspensions adjusted to pH 8.2 were thicker than the mayonnaise sample while those adjusted to pH 4.0 were thinner. All were more viscous than two commercial salad dressings.

This emulsification behavior apparently cannot be attributed solely to nitrogen-solubility patterns (Figure 8), which have been characterized by Okaka and Potter (37). Though nitrogen-solubility profiles of cowpea proteins are influenced by pH and salt concentration changes, these effects were not clearly shown in emulsion capacity behavior (27). Emulsion-forming and thickening properties of cowpeas may also be influenced by their high carbohydrate content (about 60%, dry wt basis); the starch component is hydrophilic and may interact with protein and other components in the system. Sefa-Dedeh and Stanley (38) found that changing the ionic strength of the extracting buffer produced more profound changes in the nitrogen solubility of cowpea proteins than heating at temperatures ranging from 25 to 100°C or at 100°C for 90 min. Okaka and Potter (37) found no difference in emulsion capacity of cowpea powders prepared from peas that had been soaked in water at pH 2, 4, or 6, then steam blanched, and drum dried to reduce beany flavor.

Table I.   Emulsion capacity (EC) and viscosity of peanut
           flour as influenced by flour concentration,
           particle size, and rate of mixing.

Treatment		EC (ml oil/g flour)	Viscosity (poise)
Flour concentration (g/50 ml)	1	50.0	0.23
	3	25.0	1.63
	5	16.0	5.83
	7	12.1	10.72
	9	10.5	19.13
Particle size (mesh)	60	16.0	5.83
	100	18.0	9.33
	200	21.0	10.86
Mixing rate (rpm)	13,000	18.0	4.89
	15,000	17.0	5.13
	18,700	16.0	5.83

from Ramanatham et al. (21)

Table II.   Effect of pH adjustment on emulsion capacity
            (EC) and viscosity of cowpea flour.

pH Adjustment	EC ml oil/g flour	Viscosity[a] cps
6.4 → 4.0	28.9	26,080
6.4	36.5	47,680
6.4 → 8.2	42.2	66,240
6.4 → 4.0 → 8.2	38.0	60,160

[a]For comparison, a commercial mayonnaise had a viscosity of
44,640 cps; two commercial salad dressings, 6,400 and 16,800
cps.

from McWatters and Cherry (27)

Sefa-Dedeh and Stanley (39) attributed decreases in solubility characteristics of cowpea proteins at low ionic strength to the formation of ionic bonds 1) within the protein molecule and 2) between adjacent protein molecules leading to the formation of aggregates. Other changes in solubility at higher ionic strength appeared to result from "salting-in" and "salting-out" effects. Data in Table III show variable response in emulsion capacity of field pea and cowpea products with changes in ionic environment. Emulsion capacity was higher when the pea products were dispersed in water than in any of the salt solutions. An inverse relationship between protein concentration and emulsion capacity is apparent at each salt concentration. Data in Table IV show that each pea product formed higher viscosity emulsions in water than in salt solution. The decrease in emulsion viscosity which occurred when salt was present in the continuous phase was greater in the field pea products than in cowpea flour and may have been caused by variations in processing conditions as well as inherent differences in the seeds.

Sosulski and Youngs (40) used an air-classification process to fractionate eight legume flours into fine and coarse fractions. The fine material was associated with the protein fraction, whereas the coarse material was primarily starch. A modification of the model system method developed by Inklaar and Fortuin (41) was used to measure oil emulsification. This procedure involves stirring a protein-containing material into water contained in a beaker, adding salt, adding a fixed amount of oil from a buret while stirring, centrifuging until the volume of separated oil does not change any further, and calculating the separated oil as a percentage of the total amount of oil added. Sosulski and Youngs (40) reported oil-emulsification capacity as the percentage of oil emulsified. Data in Table V show that chickpea and Great Northern bean had high emulsion capacity. The protein fractions had greater emulsifying capacity than starch fractions.

Satterlee et al. (42) investigated the emulsion capacity of fractions of the Great Northern bean by means of a micro-emulsifier apparatus developed by Tsai et al. (43). It used small quantities of materials and provided for exclusion of air and injection of oil at a constant rate until the emulsion breakpoint was reached. Modifications in initial oil-water-protein mixing before the beginning of the run and in blender speed were made (44). Emulsion stability was determined by preparing sausages from the bean protein products and measuring cooking losses. These authors found that the albumins had greater emulsion capacity than the globulins (Table VI). The low emulsion capacity of the globulins was reflected in the low emulsion capacity of the bean protein concentrate. The globulins showed good stability in the heated sausages used in the emulsion stability test. The high emulsion capacity of the albumins was not apparent in emulsion stability, indicating the importance of evaluating both aspects of emulsification properties.

Table III.   Emulsion capacity (EC) of pea protein products dispersed in water and salt solutions.

Pea Product	% Protein	EC (ml oil/mg protein)			
		Water	0.1M NaCl	0.5M NaCl	1.0M NaCl
Field pea protein concentrate	55.6	.224	.178	.206	.208
Field pea flour	32.4	.327	.262	.247	.279
Cowpea flour	21.0	.423	.414	.363	.391

Table IV. Viscosity of emulsions prepared from pea protein products dispersed in water and salt solutions.

Pea Product	Viscosity (cps)			
	Water	0.1M NaCl	0.5M NaCl	1.0M NaCl
Field pea protein concentrate	30,240	10,080	12,640	20,480
Field pea flour	16,480	2,560	1,440	2,560
Cowpea flour	22,880	17,600	16,960	18,880

Table V. Oil emulsification capacity of legume flour, protein (PF) and starch (SF) fractions in percent, dry basis.

	% Oil Emulsified		
	Flour	PF	SF
Chickpea	94	92	79
Pea bean	64	66	20
Northern bean	92	98	57
Fababean	47	77	24
Field pea	48	48	20
Lima bean	53	70	27
Mung bean	64	64	22
Lentil	56	58	50

from Sosulski and Youngs (40)

Table VI. Emulsion capacity (EC) and stability (ES) of various
           protein fractions from the Great Northern bean.

	EC	ES (ml released/10 g emulsified meat)	
	ml oil/100 mg protein	Fat	Water + Suspension Solids
Globulins	12.4	0.03	0.8
Albumins	26.8	0.04	1.3
BPC[a]	15.4	0.13	1.2

[a]BPC (bean protein concentrate contained 65% globulins and 35% albumins).

from Satterlee et al. (42)

Sunflower Seed. Emulsion capacity of defatted sunflower meal was investigated by Huffman et al. (45) at three pH levels (5.2, 7.0, 10.8), blender speeds (4500, 6500, 9000 rpm), and oil addition rates (30, 45, 60 ml/min). With low mixing speeds and rapid rates of oil addition, optimum emulsion capacity occurred at pH 7.0. These authors related the observed emulsification properties to protein solubility, surface area and size of oil droplets, and rate of protein film formation.

Lin et al. (46) measured the emulsion capacity of defatted sunflower seed products. Data in Table VII show that sunflower flour was superior in emulsifying capacity to all other products tested. The emulsions were in the form of fine foams and were stable during subsequent heat treatments. The diffusion-extraction processes employed to remove phenolic compounds dramatically reduced emulsion capacity, although isolating the protein improved emulsion capacity to some extent.

Canella et al. (47) evaluated the effects of succinylation and acetylation on the emulsion capacity of sunflower protein concentrate (Figure 9). These authors showed that emulsion capacity for untreated and treated concentrates was low in the acidic pH region, increased until pH 7 was reached, then decreased at higher pH levels. Chemical modification by succinylation and acetylation improved emulsion capacity of sunflower protein concentrate, with the 10% succinylation treatment producing the greatest increase. Emulsion stability, determined by a heating and centrifugation test, was also improved by these chemical treatments. Improvements in emulsifying properties were attributed to changes in protein solubility and electrical charges.

Rapeseed. Methods employed in processing of rapeseed protein products influence emulsion capacities (48, 49). Kodagoda et al. (48) showed that rapeseed protein isolates from water extracts emulsified more oil than isolates from acid or alkali extracts (Table VIII). Rapeseed isolates emulsified more oil than their concentrate counterparts. Rapeseed isolates and concentrates from acid extracts were far superior in emulsion stability to rapeseed protein products from water or alkali extracts.

Sosulski et al. (49) found that high temperature extraction procedures used to detoxify rapeseed protein products by removal of glucosinolates adversely influenced emulsion capacity (Table IX). An isolate prepared from a low glucosinolate cultivar, however, had higher emulsion capacity values than the other detoxified rapeseed products and a soy flour and concentrate also tested. The authors attributed the low emulsion capacity values of the soy products and the Tower rapeseed concentrate FRI (Food Research Institute) to their low nitrogen solubilities. Gillberg (50) demonstrated that the solubility characteristics of rapeseed protein isolates were highly sensitive to and altered by small

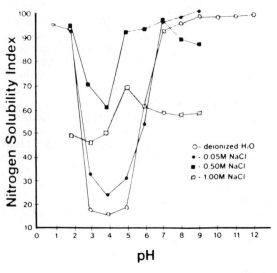

*Figure 8.   Effects of pH and salt concentration on nitrogen solubility index of raw
dehulled cowpeas (37)*

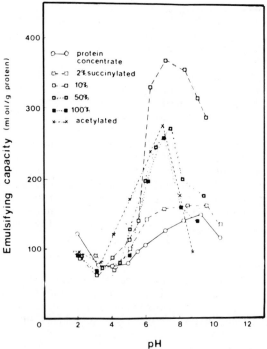

*Figure 9.   Effect of pH on emulsifying capacity of native and chemically modified
(succinylation, acetylation) sunflower seed protein concentrates (47)*

Table VII.   Emulsion capacity of defatted sunflower
seed products.

Defatted Sunflower Seed Product[a]	% Protein	% Oil Emulsified
Flour	55.5	95.1
Concentrate (DE-60)	67.9	14.0
Concentrate (DE-80)	68.3	11.3
Concentrate (DE-90)	68.0	10.1
Isolate (DE-60)	90.7	25.6

[a]Concentrates were diffusion extracted in tap water (pH 5.5) at
60°C for 4 hr, 80°C for 1.5 hr, or 90°C for 1.0 hr to remove
phenolic compounds. The isolate was prepared by additional
washing and pH adjustment of the DE-60 concentrate.

from Lin et al. (46)

Table VIII. Emulsion capacity (EC) and stability (ES) of rapeseed protein isolates and concentrates.

Product	Extraction Stage-Medium	EC ml oil/100 mg protein	ES min
Isolate	first-water	45.3	10.0
Isolate	second-acid	36.3	200.0
Isolate	third-alkali	32.4	8.0
Isolate	single-alkali	37.3	8.5
Concentrate	first-water	33.2	7.5
Concentrate	second-acid	27.5	200.0
Concentrate	third-alkali	25.1	7.0
Concentrate	single-alkali	29.5	6.5

From Kodagoda et al. (48)

Table IX. Emulsion capacity of rapeseed protein products.

Rapeseed Product	Total Gluco-sinolates mg/g	% Oil Emulsified[a]
Midas flour	17.0	20
Torch flour	7.8	29
Tower flour	1.2	46
Tower meal	1.1	34
Tower concentrate 80°	0.1	17
Tower concentrate FRI[b]	0.1	10
Tower isolate	0.0	38

[a]For comparison, a soy flour and soy concentrate emulsified 12 and 10% oil, respectively.

[b]FRI = Food Research Institute

from Sosulski et al. (49)

amounts of salts and variations in pH. Because these factors affect interaction and ionic bonding of molecules in food systems, functional properties other than solubility are also likely to be affected.

Cottonseed. Emulsion capacity and viscosity characteristics of glandless cottonseed flour dispersed in water are markedly influenced by variations in pH and flour concentration (51, 52; Figure 10). Emulsion capacity was lowest near the isoelectric pH (4.5) and highest at pH 11.5. Suspensions in the alkaline pH range emulsified more oil than those at acidic pH levels. These authors found that the soluble fraction was closely related to the viscosity of emulsions formed by glandless cottonseed flour. Viscosity values were lowest near the isoelectric pH and highest at pH 11.5. Emulsions formed by suspensions in the alkaline pH range were more viscous than those formed by acidic suspensions. This behavior may be associated with the storage globulins of cottonseed protein, most of which are soluble at alkaline pH levels (51).

Safflower. Betschart et al. (53) studied the influence of extraction, isolation, neutralization, and drying variables on emulsification properties (activity and stability) of safflower protein isolates. Emulsion activity was determined by measuring the volume of emulsified layer remaining after centrifugation of a protein isolate-water-oil mixture. Emulsion stability was determined by measuring the amount of emulsified material remaining after heating at 70-74°C for 30 min. Data in Table X show that neutralization of safflower protein isolates to pH 7 was required for emulsion formation and stabilization and was more influential than precipitation pH or drying method. The method used to dry safflower protein isolates had little influence on emulsification properties. The neutralized safflower protein products performed equally as well as a soy protein concentrate and isolate also tested. Thus, pH is a major contributing factor to emulsification properties of safflower protein isolates.

Vegetable Proteins as Emulsifiers in Food Systems

The general class of food-grade emulsifiers is broad, but these agents may be grouped according to the function they perform (2, 5) including:

1) reduction of surface tension at oil-water interfaces, which promotes emulsion formation, phase equilibria between oil-water-emulsifier at the interface, and emulsion stability (e.g., the action of egg yolk and mustard to emulsify oil and water in mayonnaise);
2) interactions with starch and protein components in foods that modify texture and rheological properties

Table X. Emulsification properties of safflower protein
isolates as influenced by pH and drying method.

pH Adjustment	Drying Method[a]	Emulsification Properties	
		Activity (%)	Stability (%)
9 → 5	FD	0	0
9 → 5	SD	0	0
9 → 5 → 7	FD	63 ± 1.5	62 ± 1.4
9 → 5 → 7	SD	59 ± 0.8	59 ± 1.0
9 → 6	FD	0	0
9 → 6	SD	0	0
9 → 6 → 7	FD	58 ± 2.8	60 ± 2.3
9 → 6 → 7	SD	52 ± 1.7	51 ± 0.8

[a]FD = Freeze dried; SD = Spray dried.

from Betschart et al. (53)

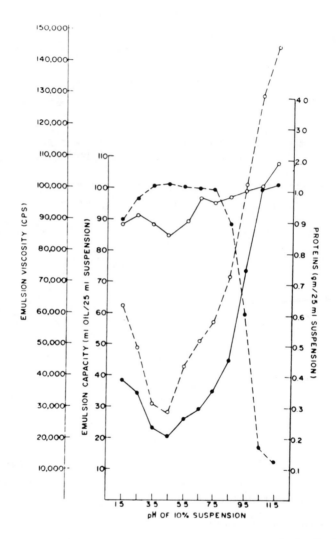

Figure 10.   Solubility and emulsifying properties of cottonseed proteins at various pH values: (○ — ○) emulsion capacity; (○ – – – ○) emulsion viscosity; (● — ●) soluble protein fraction; (● – – – ●) insoluble protein fraction (51).

(e.g., the action of mono- and di-glycerides and
stearoyl lactylates as dough conditioners and anti-
staling agents in bakery products);
3) control of crystallization in fat and sugar systems
(e.g., the action of mono- and di-glycerides in peanut
butter to prevent oil separation, or sorbitan mono-
stearate to prevent "bloom" or migration of fat to
the surface of chocolate).

In the literature, reports of using vegetable proteins, other than
soy, specifically for these functions are sparse. Vegetable
proteins have been used in breads and bakery products, not as
emulsifying agents or dough conditioners, but for protein enrichment (42, 48, 53-62). Instead of conditioning the dough and improving bread quality, however, the opposite effect is produced
when vegetable proteins are used as partial replacements for wheat
flour. Because the gluten component is diluted in composite flour
mixtures of this type, modifications in formulation, mixing, or
baking procedure are generally required to offset adverse effects
on loaf volume, crumb structure, or sensory attributes.
A few reports in the literature describe the effects of using
peanut protein in comminuted meat systems such as meat loaves (63,
64), frankfurters (65), and ground beef patties (66). In some
instances, peanut protein either produced beneficial effects
(e.g., increased tenderness and cohesiveness, 66) or no adverse
effects from a sensory, physical, or microbial standpoint (65).
In other instances, however, no benefits such as reduced shrinkage
or increased fat and water binding were derived by use of peanut
protein (63, 64). Torgersen and Toledo (64) found that protein
solubility had a negative influence on the desirable properties
of a comminuted meat system, with the more soluble proteins allowing more fat release on cooking. These authors suggested that
a profile of solubility change with temperature seemed to be a
better index of functionality than solubility at any single temperature; they concluded that the most desirable proteins for use
as additives in comminuted meats were those that possessed the
combined qualities of good gel strength, high water-absorption
capacities at 90°C, and increasing solubilities with increasing
temperatures. The mechanisms involved in meat systems, therefore,
are much more complex than emulsification alone and seem to be a
combination of emulsification, binding, and gelation.
Vaisey et al. (67) evaluated the performance of fababean and
field pea concentrates as ground beef extenders. Sensory attributes of broiled meat patties containing the protein concentrates
were more acceptable to taste panelists when these products were
added as texturized flakes than as flour. The extended products
had significantly less drip loss than all-meat controls but had
softer textures. Texturizing the concentrates by drum-drying
and flaking reduced their fat retention properties but produced
patties more like the all-beef control.

Lin et al. (68) investigated the effects of supplementing wieners with sunflower flour and diffusion-extracted protein concentrates. The sunflower products showed good emulsion stability, less shrinkage, and less cooking loss than all-meat controls. Though the inclusion of the sunflower seed products in wieners resulted in slightly lower organoleptic scores, most of the supplemented products were considered to be acceptable. These authors emphasized that the true value of vegetable proteins can only be determined by incorporating them into actual food products.

Conclusions

Most of the information pertaining to emulsification properties of vegetable proteins, other than soy, has been derived by means of model-system studies. For the most part, the application of this information to performance in actual food systems has yet to be determined. Difficulties in obtaining meaningful data regarding emulsification mechanisms of vegetable proteins stem from the lack of standardized model system tests that can simulate the processing conditions and ingredients involved in complex food systems. The information that has been obtained should not be discounted, however, but rather should be considered as a point of departure and as part of a technological base concerning physical and functional characteristics of vegetable proteins. The application of this data base to food products presents a substantial challenge to the food science and technology discipline.

Literature Cited

1. Lissant, K. J. In "Emulsions and Emulsion Technology,"
       Part I; Lissant, K. J., Ed. Marcel Dekker, Inc.:
       New York, N. Y., 1974; p. 1.

2. Lynch, M. J.; Griffin, W. C. In "Emulsions and Emulsion
       Technology," Part I; Lissant, K. J., Ed. Marcel
       Dekker, Inc.: New York, N. Y., 1974; p. 249.

3. Sumner, C. G. "The Theory of Emulsions and Their Technical
       Treatment"; The Blakiston Co., Inc.: New York, N. Y.,
       1954; p. 1.

4. Becher, P. "Emulsions: Theory and Practice"; Reinhold
       Publishing Corp.: New York, N. Y., 1965; p. 1.

5. Friberg, S. In "Food Emulsions"; Friberg, S., Ed. Marcel
       Dekker, Inc.: New York, N. Y., 1976; p. 1.

6. Stutz, R. L.; Del Vecchio, A. J.; Tenney, R. J. Food
       Prod. Devel., 1973, 7(8),52.

7. Petrowski, G. E. Adv. Food Res., 1976, 22,309.

8. Wolf, W. J.; Cowan, J. C. CRC Critical Reviews in Food
    Technol., 1971, 2(1),81.

9. Kinsella, J. E. CRC Critical Reviews in Food Sci. Nutr.,
    1976, 7(3),219.

10. Kinsella, J. E. J. Amer. Oil Chem. Soc., 1979, 56,242.

11. Hutton, C. W.; Campbell, A. M. J. Food Sci., 1977, 42,457.

12. Crenwelge, D. D.; Dill, C. W.; Tybor, P. T.; Landmann,
    W. A. J. Food Sci., 1974, 39,175.

13. Volkert, M. A.; Klein, B. P. J. Food Sci., 1979, 44,93.

14. Zakaria, F.; McFeeters, R. F. Lebensm. -Wiss. u. -Technol.,
    1978, 11,42.

15. Tornberg, E. In "Functionality and Protein Structure";
    Pour-El, A., Ed. Amer. Chem. Soc. Symposium Series 92,
    Washington, D.C., 1979; p. 105.

16. Swift, C. E.; Lockett, C.; Fryar, A. J. Food Technol.,
    1961, 15,468.

17. Saffle, R. L. Adv. Food Res., 1968, 16,105.

18. Tornberg, E.; Hermansson, A.-M. J. Food Sci., 1977, 42,468.

19. Pearce, K. N.; Kinsella, J. E. J. Agric. Food Chem., 1978,
    26,716.

20. Puski, G. Cereal Chem., 1976, 53,650.

21. Ramanatham, G.; Ran, L. H.; Urs, L. N. J. Food Sci.,
    1978, 43,1270.

22. McWatters, K. H.; Holmes, M. R. J. Food Sci., 1979, 44,765.

23. Carpenter, J. A.; Saffle, R. L. J. Food Sci., 1964, 29,774.

24. Ivey, F. J.; Webb, N. B.; Jones, V. A. Food Technol., 1970,
    24,1279.

25. Rhee, K. C.; Cater, C. M.; Mattil, K. F. J. Food Sci.,
    1972, 37,90.

26. McWatters, K. H.; Cherry, J. P.; Holmes, M. R. J. Agric.
    Food Chem., 1976, 24,517.

27. McWatters, K. H.; Cherry, J. P. J. Food Sci., 1977, 42,1444.

28. Sundar, R. S.; Rao, D. R. Lebensm. -Wiss. u. -Technol.,
    1978, 11,188.

29. Beuchat, L. R. J. Agric. Food Chem., 1977, 25,258.

30. Beuchat, L. R.; Cherry, J. P.; Quinn, M. R. J. Agric.
    Food Chem., 1975, 23,616.

31. Sekul, A. A.; Vinnett, C. H.; Ory, R. L. J. Agric. Food
    Chem., 1978, 26,855.

32. Beuchat, L. R. Lebensm. -Wiss. u. -Technol., 1977, 10,78.

33. Quinn, M. R.; Beuchat, L. R. J. Food Sci., 1975, 40,475.

34. McWatters, K. H.; Cherry, J. P. J. Food Sci., 1975, 40,1205.

35. Cherry, J. P.; McWatters, K. H.; Holmes, M. R. J. Food
    Sci., 1975, 40,1199.

36. McWatters, K. H.; Holmes, M. R. J. Food Sci., 1979, 44,774.

37. Okaka, J. C.; Potter, N. N. J. Food Sci., 1979, 44,1235.

38. Sefa-Dedeh, S.; Stanley, D. J. Agric. Food Chem., 1979,
    27,1244.

39. Sefa-Dedeh, S.; Stanley, D. J. Agric. Food Chem., 1979,
    27,1238.

40. Sosulski, F.; Youngs, C. G. J. Amer. Oil Chem. Soc., 1979,
    56,292.

41. Inklaar, P. A.; Fortuin, J. Food Technol., 1969, 23,103.

42. Satterlee, L. D.; Bembers, M.; Kendrick, J. G. J. Food
    Sci., 1975, 40,81.

43. Tsai, R. Y. T.; Cassens, R. G.; Briskey, E. J. J. Food
    Sci., 1970, 35,299.

44. Satterlee, L. D.; Zachariah, N. Y.; Levin, E. J. Food
    Sci., 1973, 38,268.

45. Huffman, V. L.; Lee, C. K.; Burns, E. E. J. Food Sci.,
    1975, 40,70.

46. Lin, M. J. Y.; Humbert, E. S.; Sosulski, F. W. J. Food
    Sci., 1974, 39,368.

47. Canella, M.; Castriotta, G.; Bernardi, A. Lebensm.
    -Wiss. u. -Technol., 1979, 12,95.

48. Kodagoda, L. P.; Nakai, S.; Powrie, W. D. Can. Inst.
    Food Sci. Technol. J., 1973, 6,266.

49. Sosulski, F.; Humbert, E. S.; Bui, K.; Jones, J. D.
    J. Food Sci., 1976, 41,1349.

50. Gillberg, L. J. Food Sci., 1978, 43,1219.

51. Cherry, J. P.; McWatters, K. H.; Beuchat, L. R. In
    "Functionality and Protein Structure"; Pour-El, A., Ed.
    Amer. Chem. Soc. Symposium Series 92, Washington, D.C.,
    1979; p. 1.

52. Berardi, L. C.; Cherry, J. P. Cotton Gin and Oil Mill Press,
    1979, 80,14.

53. Betschart, A. A.; Fong, R. Y.; Hanamoto, M. M. J. Food
    Sci., 1979, 44,1022.

54. Okaka, J. C.; Potter, N. N. J. Food Sci., 1977, 42,828.

55. Okaka, J. C.; Potter, N. N. J. Food Sci., 1979, 44,1539.

56. Beuchat, L. R. Cereal Chem., 1977, 54,405.

57. Khan, M. N.; Rhee, K. C.; Rooney, L. W.; Cater, C. M.
    J. Food Sci., 1975, 40,580.

58. Khan, M. N.; Mulsow, D.; Rhee, K. C.; Rooney, L. W.
    J. Food Sci., 1978, 43,1334.

59. Lawhon, J. T.; Cater, C. M.; Mattil, K. F. Food Prod.
    Develop., 1975, 9(4),110.

60. Rooney, L. W.; Gustafson, C. B.; Clark, S. P.; Cater,
    C. M. J. Food Sci., 1972, 37,14.

61. Matthews, R. H.; Sharpe, E. J.; Clark, W. M. Cereal
    Chem., 1970, 47,181.

62. McWatters, K. H. Cereal Chem., 1978, 55,853.

63. Hwang, P. A.; Carpenter, J. A. J. Food Sci., 1975, 40,741.

64. Torgersen, H.; Toledo, R. T. J. Food Sci., 1977, 42,1615.

65. Joseph, A. L.; Berry, B. W.; Wells, L. H.; Wagner, S. B.;
     Maga, J. A.; Kylen, A. M. Peanut Sci., 1978, 5,61.

66. Cross, H. R.; Nichols, J. E. Peanut Sci., 1979, 6,115.

67. Vaisey, M.; Tassos, L.; McDonald, B. E.; Youngs, C. G.
     Can. Inst. Food Sci. Technol. J., 1975, 8,74.

68. Lin, M. J. Y.; Humbert, E. S.; Sosulski, F. W.; Card,
     J. W. Can. Inst. Food Sci. Technol. J. 1975, 8,97.

RECEIVED September 5, 1980.

12

Nutrient Bioavailability

D. B. THOMPSON and J. W. ERDMAN, JR.

Department of Food Science, University of Illinois, Urbana, IL 61801

Functional properties, as considered from the point of view of the food processor, are those properties which impart desired physical characteristics to the products. For example, foam stability would be an important functional property to a producer of whipped toppings. From the perspective of the consumer, this functional property may contribute to satisfaction and lead to repeat purchases. Increasingly, however, the consumer is concerned with the nutritional impact of his purchases. Thus, the food industry will be pressured to expand its concept of functional properties to include nutritional considerations. Martinez (1) has recently suggested that functionality be defined as "the set of properties that contributes to the desired color, flavor, texture, and nutritive value of a product". In order to assess the nutritive value of a product, one must evaluate more than the presence of the nutrients; one must evaluate nutrient bioavailability.

Bioavailability can be defined as the extent to which a chemically present nutrient can be utilized by animals (humans) (2). Bioavailability can be influenced directly or indirectly by many physiological, pathological, chemical, nutritional, and processing conditions. Discussion in this chapter will be limited to unit food processing effects upon the bioavailability of nutrients from plant protein foods. The bioavailability of amino acids, carbohydrates, lipids, vitamins and minerals from processed foods will be selectively reviewed.

Amino Acids

Experimental Procedures. Nutrient bioavailability is a complex subject when applied to protein components. In fact, it is an error to speak of the bioavailability of the protein; rather one must consider the availability of the individual amino acids which make up the protein. Amino acid availability implies sufficient digestion of the protein in the intestines to allow absorption into the tissue. Then, in the case where

0097–6156/81/0147–0243$08.00/0
© 1981 American Chemical Society

an amino acid is limiting, there should be no restriction to
utilization of that amino acid. When these criteria are
satisfied, the amino acid may be termed available. One may
chemically analyze the constituent amino acids in a protein,
but this analysis is best viewed as an estimate of the
potential nutritional value of the protein. If digestion,
absorption, and utilization are unhindered, then the potential
may be realized. However, these processes are usually
inefficient to some degree, leading to the important
consideration of amino acid availability. The subject is
complicated in that the extent of the inefficiency varies with
the animal, other dietary components, and the treatment to
which the protein may have been subjected.

The influence of processing on amino acid availability may
be profound, especially under extreme conditions. Processing
can increase amino acid availability, often by increasing
digestibility or inactivating antinutritional factors.
Processing may also serve to decrease amino acid availability.

Protein hydrolysis in 6N HCl and subsequent analysis to
determine amino acids (except tryptophan, which is acid labile)
chemically present is a first step in protein quality evaluation. The chemical score and the EAA index represent attempts
to use this information to chemically estimate nutritional
quality of protein; their obvious limitation is their disregard
for amino acid availability. The chemical score is obtained by
evaluating the percent of the limiting amino acid in comparison
to that amino acid in whole egg protein (3). The EAA index is
the geometric mean of the ratios of each of the essential amino
acids to those amino acids occurring in whole egg (4).

Since lysine is often the limiting amino acid in plant
protein and is especially sensitive to processing damage, this
amino acid has received much attention. Carpenter (5) has
devised a method of determining "available lysine" using
dinitrofluorobenzene (DNFB). On the assumption that a free
ε-amino group will represent an undamaged, unmodified lysine
residue, Carpenter measured available lysine by the extent to
which DNFB reacts with free ε-amino groups. While this
approach has considerable merit in assessing damage to lysine,
the word available is open to question. Some of Carpenter's
"available lysine" may not be absorbed (6); thus the residue
would not be truly available to the organism in vivo.

Further attempts to evaluate nutritional quality of
protein have been made both in vivo and in vitro. In vivo
measurements have been performed with humans, experimental
animals (often rats), and with lower organisms such as
Tetrahymena pyriformis. Experimental design is simpler using
the lower organisms; at the same time, the strength of the
conclusions that may be extended to humans is lessened. There
has been much discussion regarding the type of utilization
model to be determined. In the U.S., the legally acceptable

model is the protein efficiency ration (PER):

$$PER = \frac{weight\ gain}{protein\ consumed}$$

which measures utilization for growth. Biological value (BV):

$$BV = \frac{nitrogen\ retained}{nitrogen\ absorbed}$$

is often determined, accounting for nitrogen utilization for both maintenance and growth, and allowing use of either human or animal subjects. As with the chemical score approach, a serious limitation of these feeding studies is that the only amino acid for which availability can be determined is the limiting amino acid in the test diet. This limiting amino acid may or may not be of practical importance in a normal mixed diet.

In vitro attempts have been made to simulate digestion in order to measure the digestibility of a protein by digestive enzymes. The pepsin-pancreatin digest index is just one example of an enzyme system which may be applied to a test protein prior to analysis of released amino acids. These digestibility values are appropriate to the extent that they simulate the digestive tract, the subtle workings of which are not easily imitated. The greatest value of this approach is not for accuracy but for a sensitive method of monitoring changes in the protein due to treatment (7). In addition, in vitro work may often provide clues regarding the reason protein quality is improved or reduced.

A hybrid approach combining in vitro and in vivo methods is the everted sac technique of Wilson and Wiseman (8), which has been used to estimate both digestibility and absorption. This method uses the rat small intestine outside of the animal.

Processes Affecting Nutritional Value of Amino Acids. Many unit processing operations have been shown to influence protein quality. Heat, commonly applied to protein to increase digestibility, can not only make the protein intrinsically more digestible but can inactivate inhibitors to protein digestion. Denaturation of protein is thought to be the mechanism for the increased digestibility and the inactivation of inhibitory substances.

Excessive heat can lead to decreased digestibility and absorption, resulting in reduced amino acid availability. A number of mechanisms involving cross-link formation have been suggested to account for these changes. The Maillard reaction involves the complexation of a reducing sugar and a free amino group of the protein, generally the ε-amino group of lysine. The Amadori compound which forms in the preliminary stages of

Maillard browning will yield the constituent amino acid upon 6N HCl hydrolysis and analysis, but the Amadori amino acid moiety may be unavailable to the rat (9) since intestinal hydrolysis is less harsh than 6N HCl hydrolysis. Thus, we have a clear distinction between potential nutritional value and actual availability.

Another type of cross-link formation resulting from excessive heat is the formation of isopeptides, in which the ε-amino group of lysine is involved in a new peptide bond, probably by displacing the amide group of asparagine or glutamine. The lysine moiety of free ε -N-(β-aspartyl)-lysine was found to be only very slightly available to the rat (10), whereas the lysine moiety of free ε(γ-L-glutamyl)-L-lysine was determined to be almost completely available (10,11). It has been hypothesized that the isopeptide itself may be absorbed and subsequently hydrolyzed by a kidney ε-lysine deacylase, and that only after hydrolysis could the lysine be utilized.

It now appears more likely that gut enzymes may be responsible for partial isopeptide hydrolysis (12). Waibel and Carpenter (11) suggest that hydrolysis may occur within the intestinal wall after absorption of the isopeptide. However, these studies on absorption of the free ε(γ-L-glutamyl)-L-lysine may not be relevant to cross-links in an overheated protein.

It may be concluded that cross-linking due to isopeptide formation is probably responsible for decreased overall digestibility of the protein (6,12); thus, if the digestive enzymes are not able to release the smaller peptides, then the question of availability of the isopeptides per se is beside the point. Ford (13) points out that the rate of digestibility of isopeptide links may preclude the maximum availability of the lysine as a result of the lysine entering the system too late to be effectively utilized by the tissue. Thus, the lysine in these linkages would be at least partly unavailable. Other workers have shown that small quantities of isopeptides are found in the urine (14).

Excessive heat can cause destruction of amino acid residues. The amino acid most susceptible to direct heat destruction is cystine. Although not an essential amino acid, cystine does have a sparing effect on the dietary requirement for methionine. As a result, cystine destruction can be nutritionally important. In addition, many vegetable proteins are limiting in the sulfur amino acids. Cystine destruction would be particularly harmful for these proteins.

Excessive heat due to roasting can cause racemization of amino acid residues (15). Most amino acids are only available in the L form. Consequently, complete racemization could be equivalent to a 50% decrease in availability for the residues affected.

Alkali treatment of protein can be used to increase

solubilization, for the production of isolates, and in the spinning process for texturization. This treatment has important benefits in terms of some functional properties, but it must be applied with consideration for amino acid availability. Alkali treatment, if excessive, can result in amino acid destruction, especially to cystine. It can also cause racemization. Cross-links may form by reaction with dehydroalanine, a degradation product of cystine or serine. Lysinoalanine (LAL) and lanthionine are two of the possible reaction products.

Considerable attention has been given to LAL, which has been shown to be nephrotoxic to the rat (16), the only species so far shown to exhibit sensitivity to LAL. De Groot et al. (17) found that free LAL is more nephrotoxic than protein-bound LAL. The difference may be a beneficial consequence of reduced digestibility of protein containing LAL. Woodard and Short (16) found both protein-bound LAL and free LAL to be nephrotoxic. The disagreeing results from these two laboratories have not been satisfactorily explained, but Struthers et al. (18) described data which suggests that differences in the strains of rats may be responsible. Karayiannis et al. (19) claimed that differences in diets, not the strain of rats, were responsible for variable nephrotoxicity resulting from feeding protein-bound LAL. They suggested that other nutritional factors modulated development of kidney lesions, and that the balance of essential amino acids in the diet may have been an important factor. Gould and MacGregor (20) and Struthers et al. (18) have also reviewed the controversy regarding biological effects of LAL. Not only do species respond differently to LAL, but the extent of LAL formation varies with the type of protein treated (21). Damage due to alkali treatment is induced and/or exacerbated by heat.

The availability of amino acids from protein may be affected by the refinement of the types of protein molecules. In some protein sources storage protein may be isolated from non-storage protein, or acid-precipitated proteins isolated from whey proteins. In the preparation of classical protein isolates by solubilization and acid precipitation, the amino acid make-up of the isolate differs from that of the extract. Consequently, consideration must be given to the amino acids chemically present in a refined protein.

Another method of protein treatment involves enzymatic modification. Enzymes per se may be added or the protein may be subjected to the microbial enzymes produced from fermentation. There is little evidence to indicate that the amino acid availability is influenced by these treatments, although one might expect benefits from increased digestibility. The plastein reaction is a type of enzymatic modification, in which protein hydrolysis is followed by resynthesis of peptide bonds. Evidence indicates that amino

acid availability is not influenced by this treatment (22).
   A newer method of protein treatment is chemical
modification of the amino acid residues, especially lysine.
Acylation of the ε-amino group has profound effects upon the
protein structure due to elimination of the cationic amino
group. Succinylation has even more profound effects since the
amino group is essentially replaced by an anionic carboxyl
group. Functional properties have been shown to be greatly
improved by acylation (23), but lysine availability after
modification is a matter of concern. Bjarnason and Carpenter
(12) have shown free ε-N-acetyl lysine to have only 50% of the
availability of lysine and free ε-N-propionyl lysine to have no
lysine availability. Since Leclerc and Benoiton (24) have
shown N-acetyl lysine but not N-propionyl lysine to be a
substrate for rat kidney ε-lysine deacylase, these availability
studies suggest that the activity of ε-lysine deacylase in the
kidney may determine the availability of lysine in acylated
protein. ε-N-succinyl lysine has not been evaluated as a
substrate for ε-lysine deacylase. Finot et al. (10) found no
hydrolysis of ε-N-leucyl-lysine nor ε-N-palmityl-lysine in
homogenates of rat intestinal mucosa, liver, or kidney. Thus,
simple ε-N-acyl-lysine with acyl groups greater than two
carbons would not appear to be available. However, Padayatti
and Van Kley (25) have shown that an ε-peptidase is capable of
hydrolyzing ε-N-(α-aspartyl)-lysine and ε-N-leucyl-lysine; thus
these derivatives are probably available. Since ε-N-succinyl
lysine is structurally similar to both ε-N-butyryl-lysine and
ε-N-(α-aspartyl)-lysine, it is difficult to predict the
availability of the lysine in ε-N-succinyl lysine.
ε-N-propionyl lysine from lactalbumin is somewhat available
(Table I), suggesting that the longer action of gut enzymes may
result in some significant hydrolysis (12).

TABLE I
EFFECT OF ACYLATION ON LYSINE AVAILABILITY[1]

	Estimated % Activity[2] of Lysine
ε-acetyl-L-lysine	50
ε-propionyl-L-lysine	0
Bovine plasma albumin	85
Acetyl BPA	67
Lactalbumin	122
Formyl lactalbumin	77
Propionyl (40%) lactalbumin	107
Propionyl (95%) lactalbumin	43

[1] From (12).
[2] Compared to L-lysine hydrochloride = 100.

Conclusions and application of conclusions regarding amino acid availability must be kept in perspective. A change in availability may have little practical effect if the protein in question is not a significant dietary source of protein. If the limiting amino acid in the protein is not the limiting amino acid in the diet, a change in availability could be equally unimportant. In fact, if proteins are fed in complementary fashion, the limiting amino acid in the mixed diet may be different than the limiting amino acid in either protein source fed as the sole source. In addition, the limiting amino acid depends upon the animal used in the test. Thus, one must carefully consider the practical relevance of amino acid availability determinations.

Processing of Soy foods and Amino Acid Availability.

Soybeans are an excellent vegetable source of lysine. The first limiting amino acid is methionine. Despite its generally favorable amino acid profile, unheated soy protein is an undesirable source of amino acids due to poor digestibility. This poor digestibility has often been ascribed to trypsin inhibitors (TI), which are readily heat-denatured proteins. However, Liener (26) found no correlation between the level of TI activity and PER. His in vitro studies showed that only about 40% of the difference in digestibility between unheated and optimally heated soy extract was due to TI activity; the rest was attributable to the protein being in the undenatured state. Even fully denatured legumenous protein is incompletely attacked by digestive enzymes, and the availability of sulfur amino acids may be quite low.

Heat treatment is the preferred method of reducing trypsin inhibitor activity and increasing digestibility. Several investigators have demonstrated a delicate balance between the heat treatment necessary to increase digestibility and reduce trypsin inhibitor activity and the excessive heat which can reduce digestibility and with it the amino acid availability. Hackler et al. (27) investigated the effect of heating soy milk at 121°C for varying periods. They showed that the optimum time of heating was 4 to 8 minutes. Longer times decreased the PER. Wing and Alexander (28) found that a maximum PER for soybean meal was achieved with microwave heating for 2.5 to 3.0 minutes; the PER decreased dramatically at 5.0 minutes heating.

Iriarte and Barnes (29) showed that cystine destruction by heat made this amino acid first limiting for the rat. However, cystine supplementation did not return the nutritional value to the optimum level. These workers were unable to determine the second limiting amino acid and could not rule out the possible development of a toxicity factor. Taira (30) determined that only cystine was destroyed under heating conditions commonly

employed in soybean processing. He also found (31) that
cystine destruction was independent of moisture content.

Kellor (32) points out that overheating adversely affects
palatability. Precise control of heat for desolventizing and
in subsequent processing steps is needed. Stott and Smith (33)
showed no negative effects to lysine or methionine for a
typical commercial desolventizing-toasting process.

Soy protein is refined in production of concentrates and
isolates. The process of refinement of these two products is
considerably different (34). Concentrates are produced by
leaching defatted flakes or flour to remove most of the sugars
and ash, leaving a protein and polysaccharide mixture of about
70% protein. Isolates are classically prepared by alkali
extraction of defatted meal, followed by precipitation of
protein curd at the isoelectric point. Washing of the curd
produces an isolate of at least 90% protein. Although isolates
are a higher percentage of protein, relatively more of the
original seed protein will be found in concentrates.
Isoelectric precipitation results in loss of whey protein as
well as undesirable non-protein components. Gillberg (35) has
shown that the cystine composition of the isolate is lower than
for the meal extract because non-precipitated whey protein has
a relatively larger proportion of the total sulfur amino acids.
This observation is reinforced by Mattil (36), who showed that
the range for amino acid composition was different for several
commercial concentrates as compared to commercial isolates. He
found methionine to be lower for isolates than for
concentrates.

Mattil (36) emphasized the difference in amino acid
composition within the two classifications for commercial
products. He pointed out that some of the differences are
deliberate; a high solubility isolate could well have a
different amino acid composition than an isolate with low
solubility. Mattil evaluated isolate PER's, which ranged from
1.1 to 1.75. These values were in apparent correlation with
methionine analyses (from 0.9 to 1.2 g/16 g N).

Heat treatment of commercial isolates was studied by Cogan
et al. (37) who found that heat could be applied either before
or after isolate formation without affecting PER. However,
Longenecker et al. (38) point out apparent marked differences
in manufacturing processes regarding heat treatment of
commercial concentrates and isolates. They found that in most
cases heat treatment had apparently not been optimal and that
resulting losses in nutritional value could be quite large.
Longenecker and Lo (39) analyzed the effect of excessive heat
treatment on soy isolate and found only a 14% decrease in
methionine by chemical analysis but a PER drop from 2.10 to
1.13. Using a plasma amino acid technique, they concluded that
methionine availability was reduced 46% by the excess heat, a
finding which more reasonably explains the large PER drop.

Alkali treatment is used to solubilize and isolate proteins, to improve foaming and emulsifying properties, and to obtain protein solutions suitable for spinning fibers (17). De Groot and Slump (40) studied the influence of alkali on soy protein isolates, monitoring the production of lysinoalanine and changes in amino acid content. They found that above pH 10, treatment at 40°C for 4 hours resulted in decreased cystine and increased LAL (Figure la). They also found that at pH 12.2 for 4 hours, lysine and cystine content steadily decreased with increasing temperatures from 20° to 80°C, and LAL content increased dramatically. At pH 12.2 and 40°C they reported that the greatest loss in cystine and increase in LAL occurred in the first hour (Figure lb). Thus they concluded that exposure of soy protein isolate at pH 12.2 for only a short time would destroy some cystine and decrease the nutritive value. Decreased threonine availability (not shown) in the pH 12.2, 40°C, 4 hour samples was apparent from methionine supplementation studies. The authors speculate that racemization of threonine could be responsible. Pepsin-pancreatin digestion of the pH 12.2, 40°C, 4 hour samples showed decreased digestibility. Absorption was measured with everted intestinal sacs and was seen to vary according to the amino acid. Because LAL and nitrogen utilization showed a negative correlation, these authors suggested measurement of LAL as an estimation of alkaline processing damage in soy. In later work, de Groot et al. (17) suggested that cystine is a sensitive indicator of losses in nutritive value, while safety considerations would be related to LAL content.

Sternberg et al. (41) analyzed numerous foods and food ingredients for LAL. They found LAL to be commonly occurring at greater than 100 µg/g protein in cooked (but not raw) frankfurter, chicken and egg white. Of food ingredients tested, commercial soy protein isolate samples varied from 0-370 µg/g protein. These data indicate that while processing can increase LAL levels in soy, LAL levels in soy products may be comparable to levels encountered in everyday cooked foods. Gould and MacGregor (20) concluded that human intake of LAL is low compared to levels needed to cause kidney damage in the rat. They point out, however, that certain infant milk formulas have been shown to have LAL at 200-600 ppm, and that this food may constitute 100% of the diet. They see cause for concern, considering the unknown relative sensitivity of man to LAL.

Finley and Kohler (42) noted that oxygen is apparently required for the formation of high levels of LAL in soy isolate and sodium caseinate. They were able to control LAL formation by limiting the amount of oxygen either by mixing under nitrogen or by the addition of reducing agents. Even with severe alkali treatment (60°C, 8 hours, 0.1 N NaOH) the LAL

content was 0.003 g LAL/16 g N or below when mixed under
nitrogen after addition of bisulfite, bisulfide, or cysteine.
   Soy protein may be texturized in two ways: through fiber
spinning or thermoplastic extrusion. Fiber spinning involves
treatment of soy isolate with alkali, spinning, and coagulation
in acid, whereas extrusion subjects soy flour to high
temperature and pressure, releasing the thermoplastic material
through a die (34).
   Van Beek et al. (43) examined spun soy isolate and found
that as the sole protein source to rats it is fully capable of
supporting normal growth when methionine is added. Earlier,
Bressani et al. (44) tested spun soy products + egg albumin +
wheat gluten and found the protein slightly less adequate than
milk at lower nitrogen levels, but equivalent to milk at higher
levels. These experiments suggest that spun soy isolate may be
an excellent protein source.
   Kies and Fox (45) examined extrusion-texturized soy
protein in comparison to beef protein. At 8.0 g nitrogen per
day they found that both products met the human adult male
requirements (i.e., they gave a positive nitrogen balance),
whereas at 4.0 g nitrogen per day beef was superior to the
extended soy (i.e., it had a less negative nitrogen balance).
Methionine supplementation partially overcame the difference at
the lower level. This study shows the adequacy of protein
quality of extrusion-textured soy protein for humans fed
adequate quantities of protein. The authors point out that
applying the 6.25 conversion factor to both proteins might well
put the soy protein at a disadvantage if the true conversion
factor for soy is lower, as has been suggested.
   Mustakas et al. (46) evaluated the effects of
extruder-processing on nutritional quality, flavor, and
stability of the product in an attempt to describe extruder
conditions which would be acceptable in all three respects.
Urease activity was used as an estimation of trypsin inhibitor
activity; thus the area between the two urease curves in Figure
2 indicates processing conditions which strike a balance
between too much and too little heat treatment, showing optimal
nutritional quality. Using the flavor and peroxide value
isograms, processing conditions may be chosen such that
acceptable flavor and stability may also be achieved.
   Recently, Jeunink and Cheftel (47) have attempted to
illuminate the mechanism of extrusion texturization. They
attributed the low product solubility to new disulfide bonds
and non-covalent interactions, but they could not rule out a
contribution due to formation of isopeptide links. The small
observed increase in unavailable lysine would be consistent
with isopeptide formation.
   Acylation of soy protein has been suggested as a way of
improving its functional properties. Franzen and Kinsella (48)
studied both succinylation and acetylation of soy protein.

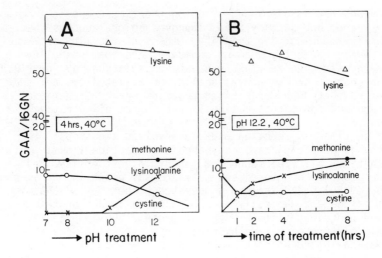

AVI Publishing Company, Inc.

Figure 1.   Effects of pH (A) and duration of alkaline treatment (B) of soy protein on levels of specific amino acids (17)

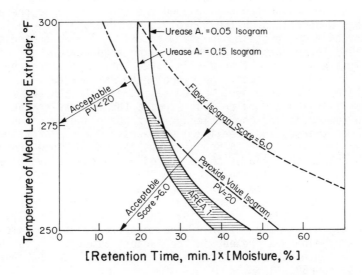

Food Technology

Figure 2.   Response contours for optimization of nutritional, flavor, and stability indicies of extruded full-fat soy flour (46)

They found negligible changes in functional properties from acetylation, but succinylation improved color, aqueous solubility, emulsifying activity, emulsion stability, emulsifying capacity, foaming capacity, and foam stability. The authors recognized that exhaustive acylation may lower PER of the protein due to unavailability of acylated lysine, and suggested that such derivatized proteins serve a functional role and not constitute a significant source of nutritive protein.

The plastein reaction usually involves two steps: hydrolysis of protein and resynthesis of peptide links. Yamashita et al. (20) found similar BV, digestibility, and weight gain for denatured soy meal and soy plastein. This same group (49,50) had described a one-step process by which amino acids may be enzymatically incorporated in intact protein to improve protein quality. With soy protein they applied a racemic mixture of D,L-methionine ethyl ester and were able to enzymatically incorporate L-methionine. As Schwimmer (51) has pointed out, one expects the methionine so incorporated to be highly available due to its location at the end of the polypeptide chains.

Kirchgessner and Steinhart (52) studied the in vitro digestion of soy protein isolate with pepsin. They showed that a relatively higher percentage of the essential amino acids threonine, valine, isoleucine, leucine and phenylalanine were found in the undigested residue and therefore would be present in lower proportions in the hydrolysate. Myers et al. (53) hydrolyzed a soy protein isolate with a fungal acid protease. They also found that the essential amino acid content of the hydrolyzate was lower than the isolate, but, as prepared by their continuous process, the hydrolyzate PER was not significantly reduced as compared to the isolate PER.

Stillings and Hackler (54) studied the effects of the tempeh fermentation, the action of Rhizopus oligosporus on dehulled, soaked, and sterilized soybeans. The PER of tempeh reached its peak after 24 hours' fermentation and then dropped. They ascribed the changes to an increase and then a decrease in tryptophan chemically present. The PER decrease after 24 hours was also attributed to an increased quantity of mold protein, which is lower in sulfur amino acid concentration than the original soy protein. Free amino acids increased markedly, presumably due to the action of the mold proteolytic enzymes. Free methionine was present in the largest amount, and as such was highly available, thus partially accounting for the initial increase in PER.

Little information is available regarding potential nutrient interaction and amino acid availability in soybeans. Bau and Debry (55) have investigated how other nutrients in a product could affect amino acid availability. Attempting to refine the protein from germinated and ungerminated soybean, they added $CaCl_2$ to form a tofu-like curd, which was then pressed and dried. The effect of alginate on the yield of protein curd was also determined. As expected (Table II), yields increased with alginate addition;

TABLE II
$CaCl_2$ PRECIPITATION OF SOY PROTEIN[1]

	% Nitrogen Recovery	PER[2]
Ungerminated protein: $CaCl_2$	59.0	2.06
Ungerminated protein: $CaCl_2$ + alginate	87.0	1.42
Germinated protein: $CaCl_2$	51.8	1.85
Germinated protein: $CaCl_2$ + alginate	82.7	1.63

[1]From (55).
[2]Corrected to casein = 2.5.

however, the PER dropped significantly with alginate addition. Supplementation with methionine (not shown) improved the alginate sample to within 0.1 on the PER scale compared to non-alginate samples which were also supplemented. The authors concluded that alginate may form enzymatically resistant linkages with methionine residues, thus decreasing methionine availability.

Wheat Products and Amino Acid Availability. As a protein source, wheat is limited by its content of lysine. Nevertheless, it is an important source of protein for humans. Processing of wheat is necessary before use, and thermal processing is often employed in some manner. Much information is available regarding the effects of milling on nutrients in wheat, but little work has been done regarding the effects of heat treatment on amino acid availability.

Hansen and co-workers have evaluated wheat protein (hard red wheat flour) heated in a model system at various temperatures (108-174°C), moisture levels (13-33%) and times (2-10 minutes) (56,57,58). They found (56) that increasing temperature or time led to increased peptide formation, and that higher moisture also gave increased peptide formation. They also found that at 108°C protein aggregation occurred, probably as a result of disulfide bond formation, and they suggested that this effect might provide an alternative to chemical oxidation for dough improvement. At higher temperatures, the protein aggregates broke down. The authors postulated that at the higher temperatures not only disulfide bonds but some peptide bonds were broken, leading to increased peptide production. Although they did not suggest it, the H_2S production noted could have resulted partly from cystine destruction which would have been followed by chain scission according to the mechanism of Hurrell and Carpenter (59).

In subsequent work with the same system, Hansen et al. (57,58) evaluated amino acid content as well as in vitro enzyme digestibility after processing under the same conditions as above. At 174° for 10 minutes they found significant losses of lysine, arginine and cysteine-cystine but not of other amino acids. They pointed out that chemical analysis indicated that cystine became limiting at these conditions. In vitro enzyme digestibility was assessed by two methods. In the first, initial rates of pepsin or trypsin digestion showed an inverse relation to processing temperature. The second method tested lysine availability by incubation with trypsin for 5 hours, followed by incubation with carboxypeptidase B for 16 hours. Availability of lysine was estimated by comparing the lysine liberated by the enzyme treatment with the lysine liberated by HCl hydrolysis (Table III). Increased lysine availability was found at 108°, but increased destruction of lysine and decreased availability of the remaining lysine was seen at the higher temperatures. Hansen et al. (57) applied the trypsin-

TABLE III
LYSINE AVAILABILITY IN HEATED WHEAT FLOUR PROTEIN[1][2]

	Control	108°C	150°C	174°C
TCB[3]	70	81	59	13
HCl[4]	100	100	92	58

[1]From (57).
[2]Expressed as HCl control = 100. 13% moisture,
 2 minutes heating.
[3]Trypsin digestion for 5 hours (37°C) followed
 by carboxypeptidase B digestion for 16 hours
 (37°C).
[4]6N HCl digestion (1°C) for 24 hours.

carboxypeptidase B technique to several commercial products and found that this technique produced a better correlation with PER than that obtained by HCl hydrolysis and lysine analysis. The authors concluded that product temperature should be kept below 125°C, and that bread would be a good vehicle for wheat protein concentrate since the internal product temperature would not exceed 100°C.

Hansen and Johnston (58) concluded that the ineffective-ness of trypsin digestion may be chiefly responsible for the lower BV of heated gluten as described by Morgan (60). They hypothesized that trypsin-resistant peptides might be poor substrates for enzymes of the intestinal mucosa. Indigestible proteins could be absorbed and excreted in the urine without utilization.

Friedman (21) studied the effect of pH on the amino acid composition of wheat gluten. At pH 10.6 and above (65°C, 3 hours) no cystine was present. LAL increased with pH above 10.6. Lysine decreased over the same range of pH's, while serine and threonine contents dropped sharply at pH 13.9. Friedman concluded that cystine is most sensitive to alkali and that LAL will form most readily if lysine residues are in proximity to the dehydroalanine formed from cystine. Thus, he explained that different steric considerations may explain the different susceptibilities of wheat gluten, casein, and lactalbumin to LAL formation.

The yeast fermentation of wheat dough prior to baking takes place at the expense of available carbohydrates. Good dough volume, texture, and structure result from the stretching of the gluten complex due to gas production. Proteolytic activity is apparently limited, especially in comparison to the tempeh fermentation previously discussed. Consequently, protein nutritive value would be much more affected by baking than by fermentation. Burrows (61) pointed out that some small improvement is probably obtained due to the comparative abundance of lysine in yeast protein, but the small proportion of yeast in the bread would make any improvement slight.

Cottonseed Processing and Amino Acid Availability. The inherent protein quality of cottonseed is good (62,63). Despite its potential quality, cottonseed is rarely used as a source of edible protein (64). Two factors discourage use of cottonseed protein. Processing conditions are often selected to maximize oil recovery at the expense of meal quality (64), and gossypol, which is harmful to monogastric animals, must be inactivated or removed (62,63). Development of glandless varieties of cottonseed minimizes the gossypol problem because gossypol is found in the pigment glands; unfortunately, only limited plantings have been made (62).

Heat treatment is often used for gossypol inactivation and in oil processing. Horn et al. (65) studied the effects of heat treatment on cottonseed protein. Because acid hydrolysis of meals subjected to different heat treatment did not reveal the vast differences determined in animal feeding experiments, these workers used in vitro enzyme digestion (pepsin, trypsin, hog mucosa in series) as a method of examining further the availability of amino acids. Very little destruction of amino acids was found as a result of processing except at 279°F for 100 minutes. The difference in acid and enzymatic hydrolysis in Table IV shows that methionine and lysine digestibility have decreased with heating, thus explaining the reduced PER values.

TABLE IV
EFFECT OF HEAT TREATMENT UPON
THE QUALITY OF COTTONSEED PROTEIN[1]

Cooking Time (min)	Max Temp (F°)	PER	Lysine Content[2]		Methionine Content[2]	
			Acid Hydrol.	Enzyme Hydrol.	Acid Hydrol.	Enzyme Hydrol.
20	234	1.83	100.0	48.0	92.0	50.9
36	230	2.01	96.6	43.3	86.7	44.6
20	261	1.73	95.3	67.7	93.3	50.0
40	261	1.70	96.4	65.7	90.0	51.8
100	279	0.52	80.7	31.9	81.3	35.7
20	200	2.26	103.4	71.7	102.0	49.1
70	200	2.01	99.4	77.2	94.0	57.1
70	180	2.16	98.2	74.0	113.3	59.8
76	240	1.69	97.4	34.6	104.0	44.6

[1] From (65).
[2] Expressed as % of unprocessed meal.

Desolventization techniques have been shown to influence lysine availability (62). Heating prior to oil removal had little effect on PER at 93°C, and only a slight effect at 108°C (66). Heat following oil removal, leads to Maillard browning unless carbohydrates are also removed before heating (62).

Fractionation of the protein has been shown to have an effect on nutritional quality (62). Results summarized in Table V show that nutritive value varies according to the

TABLE V
EFFECT OF REFINEMENT OF NUTRITIONAL QUALITY
OF COTTONSEED PROTEIN[1]

	PER	Lysine[2]	Methionine + Cystine[2]
Cottonseed			
Glandless Flour	2.42	42	26
Non-Storage Isolate	3.05	57	33
Storage Isolate	1.50	33	19
Casein	2.50	81	32
Soybean	2.19	62	24

[1]From (62).
[2]mg/g protein.

apparently different proteins which may be selectively enriched in either non-storage or storage protein isolates. Non-storage protein is of better nutritional quality due to its higher lysine content (64). Storage proteins are acid soluble and have good foam capacity and stability (64).

Sunflower Protein Amino Acid Availability.
Sunflower protein has an excellent amino acid balance except for low lysine (67), which is apparently first limiting (68). Although not recommended for use as a protein supplements for grain products, sunflower protein is excellent for supplementation of soybean (69) due to the complementary effect of the relatively high sulfur amino acid content of sunflower protein (67). Robertson (68) points out that sunflower protein is more digestible than most vegetable proteins and comparable in BV. In order to obtain the protein, the seed should be completely dehulled, which is difficult to do commercially (67). Classical procedures for isolate preparation do not remove the chlorogenic acid, and those methods which do remove chlorogenic acid are not always practical (68).

Amos et al. (70) examined the effect of heat treatment on sunflower protein. Heating of meal for 1 hour at 100°C

significantly increased rat growth as compared to no heating or
heating at 75°, 115°, or 127°C for the same time. Only slight
loss of lysine was observed by chemical analysis. Apparently,
the mild heat treatment brings about some sort of beneficial
action (such as increased protein digestibility), whereas
excess heat treatment lowers lysine availability. Heating at
100°C was seen to increase palatability as well.

The globulin fraction constitutes 70-80% of the seed
protein and has low solubility below pH 8 (69). Provansal et
al. (71) evaluated modification of sunflour protein due to the
alkaline conditions which are often used to solubilize, purify,
and obtain desired functional properties. Losses of arginine,
cystine, threonine, serine, lysine, and isoleucine were seen
with increasing severity of treatment. Because more LAL formed
than could be accounted for by cystine degradation to
dehydroalanine, degradation of serine was postulated to lead to
LAL as well. In vitro availability as measured by Pronase
digestion showed that essential amino acids were all adversely
affected as treatment increased in severity. Analysis using
L-lysine decarboxylase showed significant racemization of
lysine in 0.2M NaOH at 80°C for 1 hour, and increasing
racemization with increasing severity of treatment (Table VI).
Since D-lysine is not available to the organism, racemization
can have detrimental effects on nutritional quality.

Processing of Sesame Protein and Amino Acid Availability.
Sesame protein is high in sulfur amino acids but low in lysine
(72). As such it is an ideal protein for complementation of
legumes. Villegas et al. (73) pointed out that methionine
supplementation using sesame protein has several advantages
over supplementation with free methionine, including an even
and more adequate release into the digestive tract.

Removal of the hulls is important to reduce oxalic acid
content, increase protein content (72) and increase
digestibility (74). Sesame protein has low solubility, and one
method of increasing solubility is with partial enzymatic
hydrolysis (75). Hydrolysis with fungal enzymes increased the
soluble protein content of sesame meal from 7.21 to 64.14%,
values considered comparable to those in soy due to the tempeh
fermentation. Increased lysine in the hydrolysis extract was
accompanied by decreased methionine.

Sastry et al. (74) showed that lysine in sesame protein is
quite stable to heat. Dehulled, solvent-extracted meal
autoclaved up to 45 minutes showed no appreciable loss of
chemically available lysine, with only an 8% loss after 60
minutes. A 13% loss of available lysine was found with
screw-pressed, solvent-extracted meal autoclaved for 60
minutes.

Villegas et al. (73) studied the influence of processing
on sesame protein quality (Table VII). Sample 1 was

TABLE VI
AMINO ACID MODIFICATION OF SUNFLOWER PROTEIN ISOLATE
DUE TO ALKALINE TREATMENT[1,2]

		Treatment[3]				
	Control	0.05, 55,1	0.1 55,1	0.2 80,1	0.2 80,16	0.5 80,16	1, 80,16
Total Lysine	11	10	10	8.5	10	11	12
D-Lysine	0.4	0.5	0.5	1.3	3.7	6	6.4
1/2 Cystine	2.5	2.5	2	0	0	0	0
LAL	2.8	3	3.5	5.5	3	2	1.3

[1] From (71).
[2] mmol/100 g isolate.
[3] NaOH(m), T($^{\circ}$C), time (hr).

pre-pressed and then solvent extracted, sample 2 was treated
similarly and then sugars were extracted, and sample 3 was
subjected to an expeller operating at high pressure and
temperature, as in commercial conditions. The data in Table
VII indicate that lysine (available after enzymatic hydrolysis)
is sensitive to processing and that use of a commercial
expeller at least partially blocks lysine availability.

TABLE VII
EFFECT OF SESAME PROCESSING ON AMINO ACID AVAILABILITY[1]

Sample[2]	Lysine[3] (available to MO after enz. hydrolysis)	Lysine[3] (after acid hydrolysis)	Lysine[3] (DNFB available)	Methionine[3]
1	1.08	2.75	1.93	1.85
2	1.43	2.79	1.92	1.96
3	0.84	2.71	1.87	2.17

[1] From (73)
[2] See text for processing conditions.
[3] g/16 g N.

The most important nutritional use of sesame protein is as
a methionine source for complementation of other protein;
consequently, the low lysine content is not an important
consideration with proper complementation (69). Here we see a
case where reduced bioavailability of an amino acid is
documented but is of little practical importance in the mixed
diet in which the protein will be only a part.

Processing of Rapeseed Protein and Amino Acid Availability. Rapeseed meal contains several undesirable
components, including erucic acid, glucosinolates, and phytic
acid (76,77). At the same time, rapeseed protein is high in
lysine and sulfur amino acids and has a higher nutritional
value than any other vegetable protein (76). Plant breeding
has reduced erucic acid levels to less than 5% of the fatty
acids (77). Glucosinolates may be reduced from 4% to < 0.2%
through the formation of protein concentrates, and at this
level there is apparently little growth-depressing effect (76).
Phytic acid levels can increase from 2% to 6% during
concentrate preparation. When the concentrate is fed to
pregnant rats, anorexia may be induced unless additional zinc
is provided to make up for the zinc no longer available due to
complexation with phytate (77).
Korolczuk and Rutkowski (78) examined the amino acid
content of rapeseed isolates prepared by several methods.
Glucosinolates were not present. They found that isolates are
generally characterized by higher levels of essential amino

acids than found in the original meal. This finding is just the opposite of what is found for soybean protein isolates, where essential amino acid content is decreased upon isolate preparation.

Significance of Amino Acid Destruction. In any discussion of amino acid availability it is important not to lose sight of the practical relevance of the changes that may result from processing. Many of the above-mentioned effects have been studied in intentionally damaged protein, and it is crucial to consider the extent of similar damage in the range of commercial processing conditions. If the process does not produce a toxic compound, then one needs only to consider whether the protein will be a significant dietary protein source. One must consider whether the limiting amino acid in that protein (the only amino acid whose availability is measured) may be limiting in the diet as a whole. If it is limiting in the diet, then the issue of availability must be faced squarely.

Carbohydrate

Chemical modification of simple sugars during drying, baking, or roasting operations can either have a desirable or undesirable effect upon the organoleptic quality of the final product. We have become accustomed to the characteristic roasted or baked flavors of coffee, peanuts, popcorn, and freshly-baked bread. The color and flavor and aroma of caramel make it a useful additive for the food industry. On the other hand, the burnt flavor of overheated dry beans or soy milk reduces marketability of these products.

The Maillard browning and the high-temperature-induced caramelization reactions that produce flavor and color changes also result in the formation of non-digestible carbohydrate materials. As previously noted, the Maillard reaction can severely reduce the bioavailability of lysine and other amino acids. Free amino groups combine the reducing sugars such as lactose or glucose, and depending upon conditions irreversible dehydrated and cyclic and fragmented degradation products can result (5,79). Caramelization of glucose syrup results in formation of cyclic monomers and dehydrated polymeric compounds.

The resulting reduction in caloric value of browned or caramelized foods is of little significance for well-fed Americans. Loss of biologically available amino acids is of concern only for those consuming diets of marginal protein quality.

Some legumes and dry beans contain considerable quantities of oligosaccharides which cannot be adequately digested by man because we lack appropriate digestive enzymes. Cramps, diarrhea, nausea, and flatulence can result from consumption of these foods due to microbial attack on these compounds in the colon. These non-bioavailable sugars are partially responsible for the low human consumption of soybeans and certain types of dry beans by Western man. Genetic efforts to reduce these flatus factors show promise. Water-blanching of soybeans will remove some of these soluble sugars. The use of alkaline soaking and/or blanching of whole soybeans (0.5% NaHCO$_3$) has been shown to be beneficial in increasing oligosaccharide removal into the blanch water (80). Mild sodium bicarbonate treatment also improves the tenderization of the soybeans (81) and the mouth-feel of a soy beverage (82), but severely reduces thiamin content of the soy beverage (83).

Many persons from non-Eastern European-derived cultures are unable to tolerate large quantities of the milk sugar lactose because of the absence of enough of the active digestive enzyme, lactase. In these persons, consumption of lactose results in effects similar to those described for oligosaccharide consumption. Severe intolerance can result in malabsorption of all nutrients from the diet due to diarrhea.

Dietary fiber, the undigestible polysaccharide material of plant cell walls, is being lauded by some for prevention of colon cancer, diverticular disease, hypercholesterolemia, diabetes, etc. Although it is uncertain whether these claims are true, it is clear that increased dietary fiber will increase stool bulk, stool moisture content, and accelerate stool transit time (84). The specific chemical type as well as the particle size of the fiber may influence its physiological effects. For example, the important property of dietary fiber of increasing water holding capacity of the stool may well depend upon particle size of the fiber (85). Daily fecal wet and dry weight in young adult men fed low-fiber diets supplemented with 32 grams of either coarse or finely ground wheat bran were significantly greater by 14% and 7% for the coarse bran diets (86). These results imply that minimally-processed native fiber may be more effective (more biologically active) than micropulverized fibrous materials that are added to low calorie breads, for example.

Lipids

Under proper processing conditions minimal alterations in the structure and bioavailability of food lipids take place. However, excessive heat and/or improper storage conditions can result in a number of degradative changes in lipids. These chemical changes have a much greater effect upon palatability than upon nutritive value.

Lipids are highly digestible and provide both calories and essential fatty acids. Vegetable oils are generally high in polyunsaturated fatty acids, and therefore are high in essential fatty acids (linoleic, linolenic and arachidonic acids). Although polyunsaturated fatty acids can undergo peroxidation by autoxidation, via enzyme stimulation (lipoxygenase) or by chemical catalysis, the dietary requirement for essental fatty acids is so low (about 1 gram per day) that the requirement is easily met in a mixed diet (87).

A. E. Bender (87) in his review of food processing and nutrition has outlined the various chemical changes that occur during processing and storage of lipids. He points out that foods containing very small amounts of fats can also undergo oxidative and hydrolytic spoilage. Vegetables such as potatoes, spinach, and beans contain as little as 0.1 to 0.6% fat but this is sufficient to cause deterioration of flavor during storage.

Fats with peroxide values of 100 (88) cause no obvious effect when fed to rats. Peroxide values of 800 will cause loss of appetite and decreasing ability to gain weight, while values of 1200 can be quite toxic. Peroxide values in excess of 100 are rarely encountered with good commercial practices (89). Heat-induced cyclization or polymerization of fats results in quite toxic compounds when tested with rats. These materials may affect membrane function (90), liver enlargement (91) and other biological systems (92).

Commercial hydrogenation of oils rich in polyunsaturated fatty acids lowers melting points of the oil and reduces the heat and oxidative damage to the oil during processing and storage (92). During hydrogenation, some of the naturally-occuring cis-double bonds in fatty acids isomerize to trans form. Trans-fatty acids are absorbed and metabolized similarly to cis-fatty acids but their effects upon lipoprotein structure and biological membranes are still under investigation. Trans-fatty acids have different configuration than cis-fatty acids and may alter permeability characteristics and biological function of membranes (93). Trans-fatty acids can intensify essential fatty acid deficiency in experimental animals (94).

Vitamins

It is extremely difficult to predict the vitamin content of prepared food-stuffs. Much of the problem lies in the large variability in vitamin content of raw foods. Genetic, seasonal and environmental factors affect the concentration of vitamins, especially in foods of plant origin. In addition, various food processing procedures result in a large range of nutrient retention (87, 95). Major loss of vitamins occur due to heat or water leaching, while some vitamins are labile to acid or alkaline pH, oxygen or to light (87,96).

Vitamin C is the most heat labile vitamin. Both

L-ascorbic acid and its reversible-oxidized form,
L-dehydroascorbic acid, are active forms of the vitamin.
However, once the dehydro form is further oxidized to
diketogluconic acid, no vitamin C activity is retained. The
thermal half life of dehydroascorbic acid is less than 1 minute
at 100°C (pH 6), and 2 minutes at 70°C (87). Vitamin C is also
easily leached into cooking water and can be destroyed by a
series of plant oxidizing enzymes.

Thiamin, vitamin B_1, is the next most labile vitamin.
It is unstable to heat, neutral or alkaline pH, and can be
easily leached out of foods. Thiamin is destroyed by sulfur
dioxide (often used as a food preservative), potassium bromate
(oxidizing agent used in bread) and by sodium bicarbonate (in
chemically leavened breads).

Riboflavin and folic acid are less heat labile, but are
easily leached out of foods. Some forms of vitamin B_6 in
foods decrease in bioavailability during thermal processing
(97). Gregory and Kirk (98) investigated the bioavailability
of various forms of vitamin B_6 in simulated roasted cereal
products and found 50 to 70% loss in the forms tested. Niacin
is quite stable under all processing conditions except for
water leaching. In comparison to these water-soluble vitamins,
the fat-soluble vitamins are quite stable and only subject to
major loss from foods due to oxidation under extreme processing
or storage conditions.

Some vitamins are present in raw plant foods in "bound"
forms. Some bound forms of vitamins are biologically
unavailable while others provide partial activity. In the case
of native niacin, proper food processing can often release the
active part of the bound complex and increase the
bioavailability of this vitamin. Native niacin in many cereals
is not biologically available since it is bound to
polysaccharides and peptides (87,99). Bound forms are not
digestible by man unless the complex is hydrolyzed by alkali or
other treatment prior to consumption. In calculating the
niacin content of the diet it is not unusual to ignore niacin
from cereal sources as it is not known whether this niacin
source will become available after processing (87). Baking
with alkaline baking powder, although it destroys thiamin, will
liberate a large percentage of the niacin. Soaking corn in
lime water overnight prior to making tortillas frees bound
niacin. This practice is thought to prevent the niacin
deficiency disease, pellagra, in Central and South America.
Pellagra was prevalent in poor populations in the Sourthern
part of this country at the turn of the century. Pellagra
victims consumed large quantities of corn without the
beneficial lime treatment (100).

Much of the folacin in food is present as conjugated
pteroyl polyglutamates. Soybean contains 40% monoglutamate
analogues, but in cabbage 90% of folic acid is present in forms

with more than 5 glutamate residues (87). The intestinal enzyme folate conjugase must reduce these compounds to the monoglutamate prior to absorption. The extent of cleavage in different foods is yet unclear. Also unclear are the relative effects of heat and other treatments upon different analogues. Therefore, the absolute bioavailability of different analogues of folacin in various processed foods is not clear. The bioavailability of folates from food-stuffs and the effects of food processing upon the various analogues are an important problem which must be resolved if a reasonably accurate estimate of folacin content of foods and status in man is to be achieved (101).

Little data is available on the interaction of protein modification for improved functionality and vitamin bioavailability from modified food-stuffs. Some water-soluble and fat-soluble vitamins are protein-bound in their transport, storage and/or active forms. Therefore, methods used to cause dramatic alteration of protein conformation or chemical structure can be assumed to alter some vitamins as well.

Technologists should become aware of possible reduction of available vitamins due to such processing as some vitamins can be considered marginally deficient in American diets. This concern is especially important in those modified protein food-stuffs that can contribute a sizable portion of our daily calories.

Minerals

Minerals from plant sources are less bioavailable than from animal sources (102). Because most functionally-modified proteins are of plant origin, one must be particularly concerned about what effect plant protein modification has upon mineral bioavailability.

One of the most active areas of research on foods in the last few years has been the investigation of factors that effect human and animal mineral utilization from plant foods. Of major interest to researchers has been the bioavailability of the minerals zinc and iron, both of which are consumed in marginal levels by segments of United States population (103).

Recent review articles (2, 104-109) have described general factors that affect mineral utilization from foods. General factors such as the digestibility of the food that supplies the mineral, chemical form of the element, dietary levels of other nutrients, presence of mineral chelators, particle size of the food or supplemented minerals and food processing conditions all play a role in the ultimate mineral bioavailability (104). Many unit food processing operations can be shown to directly or indirectly alter the level or chemical form of minerals or the association of minerals with other food components.

Little agreement has been reached as to which dietary components or which food processes physiologially affect mineral availability. Many plant foods contain phytic acid, oxalic acid or other dietary fiber components that can be shown to chelate minerals. The effect of these dietary substances upon the final bioavailability of the mineral in question will depend upon the digestibility of the chelate (106).

Much of the current research has centered upon the role of phytic acid on zinc and iron bioavailability (110-124). Work performed at the authors' institution with several different types of soy foods suggests that phytic acid is a major factor affecting availability of zinc from foods derived from the legume (110-114). In addition, it appears that endogenous zinc in high-phytate foods may be a limiting factor in optimal utilization of these foods for man. We have found that fortification of soy foods (under proper conditions) with zinc, iron, magnesium, or calcium results in excellent bioavailability of these minerals. Therefore, prudent fortification should resolve the problem of poor mineral availability from some soy foods.

Other workers (115-124; for example) have also centered their efforts on the role of phytic acid on zinc and iron bioavailabiliy from both soy and wheat products. It has been suggested (120) that the phytate-to-zinc molar ratio could be used to predict zinc bioavailability in high-phytate foods. Several groups (115, 117), including ours (113), have at least partially supported this hypothesis. However, recent work from our laboratory (112) involving soy protein of similar phytate-to-zinc molar ratios clearly demonstrates that zinc bioavailability is also altered by food processing. In this study, zinc from neutralized soy concentrates and isolates was shown to be less available to the rat than was the corresponding acid-precipitated products. This is unfortunate as alkaline conditions are commonly utilized for soy and other plant proteins to obtain beneficial functional properties.

Fermentation of plant foods generally increases mineral bioavailability. Studies by Ranhotra and coworkers (121, 122; for example) have shown increased available zinc from breads and cookies that have undergone yeast fermentation. Also, iron was shown by these workers to be more highly available from unfortified breads than from breads fortified with wheat bran, soy flour or other whole grain vegetable flours.

Various processing techniques, other than fermentation, can be utilized to remove phytic acid and other dietary fiber materials that reduce mineral absorption (109). Processes would include differential extraction and filtration techniques such as ultrafiltration (125, 126, 127).

Much more work must be completed before one can adequately predict the bioavailability of minerals from foods as consumed by man. Physicochemical modification of food proteins to produce desired functional properties may alter the binding of minerals to food components and therefore alter mineral bioavailability.

Conclusions

The biological function of foods is to provide the organism with sufficient nutrients to sustain life. In some foods nutrients physically present may be unavailable to the organism for its use. Only nutrients available to the organism are termed bioavailable. Modification of functional (i.e., physicochemical) properties of plant food proteins affects both consumer preference and bioavailability of nutrients. Unit processes that modify plant food protein may affect the bioavailability of protein, non-protein constituents, or both. One often-used process for plant protein modification is heat treatment. Proper heat treatment increases the digestibility of plant protein and destroys heat-labile anti-nutritional factors. However, excess heat can cause reduced digestibility, destruction or alteration of amino acids, and loss of heat-labile vitamins. Thus heat treatment, as well as numerous other protein modification processes, is potentially beneficial as well as potential detrimental to nutritional quality. Modification of one functional property of a protein is likely to affect other functional properties, including nutritional bioavailability. For example, alkali treatment is useful for maximizing protein solubility, but it can result in formation of damaged and/or unavailable amino acid residues. Selection of processing parameters with bioavailability in mind can yield a variety of acceptable, safe and nutritious plant protein-based foods.

Literature Cited

1. Martinez, W. H. JAOCS, 1979, 56, 280.
2. Erdman, J. W., Jr. Contemporary Nutr., 1978, 3(11), 1.
3. Block, R. J.; Mitchell, H. H. Nutr. Abstr. Rev., 1946, 16, 249.
4. Oser, B. L. J. Am. Dietet. Assoc., 1951, 27, 396.
5. Carpenter, K. J. Nutr. Abst. & Rev., 1973, 43(6), 423.
6. Hurrell, R. F.; Carpenter, K. J.; Sinclair, W. J.; Otterburn, M. S; Asquith, R. S. Br. J. Nutr., 1976, 35, 383.
7. Menden, E.; Cremer, H. "Newer Methods of Nutritional Biochemistry"; Vol. IV; Albanese, A. A., Ed., Academic Press: New York, 1970.
8. Wilson, T. H.; Wiseman, G. J. Physiol., 1954, 123, 116.
9. Horn, J. M.; Linchtenstein, H.; Womack, M. J. Agr. Fd. Chem., 1968, 16, 741.

10. Finot, P.; Mottu, F.; Bujard, E.; Mauron, J. "Nutritional
    Improvement of Food and Feed Proteins"; Friedman, M.,
    Ed.; Plenum Press: New York, 1978.
11. Waibel, P. E.; Carpenter, K. J. Br. J. Nutr., 1972, 27,
    509.
12. Bjarnason, J.; Carpenter, K. J. Br. J. Nutr., 1969, 23,
    859.
13. Ford, J. E. "Proteins in Human Nutrition"; Porter, J. W. G.;
    Rolls, B. A., Eds.; Academic Press: New York, 1973.
14. Ford, J. E.; Shorrock, C. Br. J. Nutr., 1971, 26, 311.
15. Hayase, F.; Kato, H.; Funjimaki, M. J. Agr. Fd. Chem.,
    1975, 23, 491.
16. Woodard, J. C.; Short, D. D. J. Nutr., 1973, 103, 569.
17. de Groot, A. P.; Slump, P.; van Beek, L.; Feron, J. J.
    "Evaluation of Protein for Humans"; Bodwell, C. E., Ed.;
    AVI Publ. Co. Inc.: Westport, CT, 1977.
18. Struthers, B. J.; Dahlgren, R. R.; Hopkins, D. T.;
    Raymond, M. L. "Soy Protein and Human Nutrition";
    Wilke, H. L.; Hopkins, D. T.; Waggle, D. H., Eds.;
    Academic Press: New York, 1979.
19. Karayiannis, N. I.; MacGregor, S. T.; Bjeldanes, L. F.
    Fd. Cosmet. Toxicol., 1979, 17, 591.
20. Gould, D. H.; MacGregor, J. T. "Protein Crosslinking",
    Part B; Friedman, M., Ed; Plenum Press: New York, 1977.
21. Friedman, M. "Functionality and Protein Structure";
    Pour-El, A., Ed.; Amer. Chem. Soc.: Washington, D. C.,
    1979.
22. Yamashita, M.; Arai, S.; Gonda, M.; Kato, H.; Fujimaki, M.
    Agr. Biol. Chem., 1970, 34, 1333.
23. Kinsella, J. E.; Shetty, K. J. "Functionality and Protein
    Structure"; Pour-El, A., Ed.; Amer. Chem. Soc.:
    Washington, D. C., 1979.
24. Leclerc, J.; Benoiton, L. Can. J. Chem., 1968, 46, 1047.
25. Padayatti, J. D.; Van Kley, H. Biochemistry, 1966, 5,
    1394.
26. Liener, I. E. J. Fd. Sci., 1976, 41, 1076.
27. Hackler, L. R.; Van Buren, J. P.; Steinkraus, K. H.;
    El Rawi, I.; Hand, D. B. J. Fd. Sci., 1965, 30, 723.
28. Wing, R. W.; Alexander, J. C. Nutr. Rept. Int., 1971, 4,
    387.
29. Iriarte, B. J. R.; Barnes, R. H. Fd. Tech., 1966, 20,
    835.
30. Taira, H.; Taira, H.; Sugimura, K.; Sakurai, Y. Agr. Biol.
    Chem., 1965, 29, 1074.
31. Taira, H.; Taira, H.; Sugimura, K.; Sakurai, Y. Agr.
    Biol. Chem., 1965, 29, 1080.
32. Kellor, R. L. JAOCS, 1974, 51, 77A.
33. Stott, J. A.; Smith, H. Br. J. Nutr., 1966, 20, 663.
34. Horan, F. E. JAOCS, 1974, 51, 67A.

35. Gillberg, L. Nutr. Rept. Int., 1977, 16, 603.
36. Mattil, K. F. JAOCS, 1974, 51, 81A.
37. Cogan, U.; Yaron, A.; Berk, Z.; Zimmerman, G. J. Agr. Fd. Chem., 1968, 16, 196.
38. Longenecker, J. B.; Martin, W. H.; Sarett, H. P. J. Agr. Fd. Chem., 1964, 12, 411.
39. Longenecker, J. B.; Lo, G. S. "Nutrients in Processed Foods, Proteins" White, P. L.; Fletcher, D. C., Eds.; Publ. Sciences Gp. Inc.: Acton, Mass., 1974.
40. de Groot, A. P.;, Slump, P. J. Nutr., 1969, 98, 45.
41. Sternberg, M.; Yim, C. Y.; Schwende, F. J. Science, 1975, 190, 992.
42. Finley, J. W.; Kohler, G. O. Cereal Chem., 1979, 56, 130.
43. van Beek, L.; Feron, V. J.; de Groot, A. P. J. Nutr., 1974, 104, 1630.
44. Bressani, R.; Viteri, F.; Elias, L. G.; de Zaghi, S.; Alvarado, J.; Odell, A. D. J. Nutr., 1967, 93, 349.
45. Kies, C.; Fox, H. M. J. Fd. Sci., 1971, 36, 841.
46. Mustakas, G. C.; Albrecht, W. J.; Bookwalter, G. N.; McGhee, J. E.; Kwolek, W. F.; Griffin, E. L. Fd. Tech., 1970, 24, 1290.
47. Jeunink, J.; Cheftel, J. C. J. Fd. Sci., 1979, 44, 1322.
48. Franzen, K. L.; Kinsella, J. E. J. Agr. Fd. Chem., 1976, 24, 788.
49. Yamashita, M.; Arai, S.; Amano, Y.; Fujimaki, M. Agr. Biol. Chem., 1979, 43, 1065.
50. Yamashita, M.; Arai, S.; Imaizumi, Y.; Amano, Y.; Fujimaki, M. J. Agr. Fd. Chem., 1979, 27, 52.
51. Schwimmer, S. "Protein Nutritional Quality of Foods and Feeds", Part 2; Friedman, M., Ed.; Marcel Dekker, Inc.: New York, 1975.
52. Kirchgessner, M.; Steinhart, H. Z. Tierphysiol., Tirernahrg., u. Futtermittelkde., 1974, 32, 240.
53. Myers, D. V.; Ricks, E.; Myers, M. J.; Wilkinson, M.; Iacobucci, G. A. "Proceedings IV International Congress Food Science and technology", Vol. V; Instituto Nacional de Ciencin y Techologia de Alimentos: Madrid, Spain, 1974.
54. Stillings, B. R.; Hackler, L. R. J. Fd. Sci., 1965, 30, 1043.
55. Bau, H. M.; Debry, G. JAOCS, 1979, 56, 160.
56. Hansen, L. P.; Johnston, P. H.; Ferrel, R. E. Cereal Chem., 1975, 52, 459.
57. Hansen, L. P.; Johnston, P. H.; Ferrel, R. E. "Protein Nutritional Quality of Foods and Feeds"; Friedman, M., Ed.; Marcel Dekker, Inc.: New York, 1975.
58. Hansen, L. P.; Johnston, P. H. Cereal Chem., 1976, 53, 656.
59. Hurrell, R. F.; Carpenter, K. J. "Protein Crosslinking", Part B; Friedman, M., Ed.; Plenum Press: New York, 1977.

60. Morgan, A. F. J. Biol. Chem., 1931, 90, 771.
61. Burrows, S. "The Yeasts" Vol. 3; Rose, A. H.;
    Harrison, J. S., Eds.; Academic Press: New York, 1970.
62. Martinez, W. H.; Hopkins, D. T. "Protein Nutritional
    Quality of Foods and Feeds"; Friedman, M., Ed.; Marcel
    Dekker Inc.: New York, 1975.
63. Cherry, J. P.; Berardi, L. C.; Zarins, Z. M.; Wardsworth,
    J. I.; Vinnett, C. H. "Nutritional Improvement of Food
    and Feed Protein"; Friedman, M., Ed.; Plenum Press: New
    York, 1978.
64. Spadaro, J. J.; Gardner, H. K., Jr. JAOCS, 1979, 56, 422.
65. Horn, M. J.; Blum, A. E.; Womack, M.; Gersdorff, C.E.F.
    J. Nutr., 1952, 48, 231.
66. Cross, D. E.; Hopkins, D. T.; D'Aquin, E. L.; Gastrock, E.
    A. JAOCS, 1970, 47, 4A.
67. Sosulski, F.; Flemming, S. E. JAOCS, 1977, 54, 100A.
68. Robertson, J. A. Critical Rev. Fd. Sci. Nutr., 1975, 6,
    201.
69. Sosulski, F. JAOCS, 1979, 56, 438.
70. Ames, H. E.; Burdick, D.; Seerley, R. W. J. An. Sci.,
    1975, 40, 90.
71. Provansal, M. M. P.; Cuq, J-L. A.; Cheftel, J-C. J. Agr.
    Fd. Chem., 1975, 23, 938.
72. Johnson, I. A.; Suleiman, T. M.; Lusas, E. W. JAOCS,
    1979, 56, 463.
73. Villegas, A. M.; Gonzalez, A.; Calderon, R. Cereal Chem.,
    1968, 45, 379.
74. Sastry, M. C. S.; Subramanian, N.; Parpia, H. A. B.
    JAOCS, 1974, 51, 115.
75. Sriekantiah, K. R.; Ebine, H.; Ohta, T.; Nakano, M. Fd.
    Tech., 1969, 23, 1055.
76. Ohlson, R.; Anjon, K. JAOCS, 1979, 56, 431.
77. Slinger, S. J. JAOCS, 1977, 54, 94A.
78. Korolczuk, J.; Rutkowski, A. "Proceedings IV International
    Congress Food Science and Technology" Vol. V; Instituto
    Nacional de Ciencia y Techologia de Alimentos: Madrid,
    Spain, 1974.
79. Hodge, J. E.; Osman, E. M. "Principles of Food Science,
    Part I: Food Chemistry"; Fennema, O. R., Ed.; Marcel
    Dekker, Inc.: New York, 1976.
80. Ku, S.; Wei, L. S.; Nelson, A. I.; Hymowitz, T. J. Fd.
    Sci., 1976, 41, 361.
81. Nelson, A. I.; Steinberg, M. P.; Wei, L. S. J. Fd. Sci.,
    1976, 41, 57.
82. Kuntz, D. A. "Processing Factors Affecting the
    Organoleptic Quality of Illinois Soy Beverages"; Ph.D.
    Thesis; University of Illinois: Urbana, IL, 1977.
83. Bankhead, R. R.; Weingartner, K. E.; Kuntz, D. A.; Erdman,
    J. W., Jr. J. Fd. Sci., 1978, 43, 345.

84. Connell, A. M. "Nutrition in Disease: Fiber";
    Lantz, J. C., Ed.; Ross Labs.: Columbus, Ohio, 1978.
85. Kirwan, W. O.; Smith, A. N.; McConnell, A. A. Br. Med. J.,
    1974, 4, 187.
86. Heller, S. N.; Hackler, L. R.; Rivers, J. M.; Van Soest,
    P. J.; Roe, D. A.; Lewis, B. A.; Robertson, J. Am. J.
    Clin. Nutr, in press, 1980.
87. Bender, A. E. "Food Processing and Nutrition"; Academic
    Press: London, 1978.
88. Gray, J. I. JAOCS, 1978, 55, 539.
89. Dugan, L., Jr. "Principles of Food Science: Food
    Chemistry", Part I; Fennema, O. R., Ed.; Marcel Dekker,
    Inc.: New York, 1978.
90. Nixon, J. E.; Eisele, T. A.; Wales, J. H.; Sinnhuber, R. O.
    Lipids, 1974, 9, 314.
91. Iwaoka, W. T.; Perkins, E. G. Lipids, 1976, 11, 349.
92. ANON. Dairy Council Digest, 1975, 46, 31.
93. Sgoutas, D.; Kummerow, F. A. Am. J. Clin. Nutr., 1970,
    23, 1111.
94. Hill, E. G.; Johnson, S. B.; Holman, R. T. J. Nutr., 1979,
    109, 1759.
95. Erdman, J. W., Jr. Fd. Tech., 1979, 33(2), 38.
96. Harris, R. S.; Karmas, E. "Nutrition Evaluation of Food
    Processing", 2nd Ed.; AVI Publ. Co. Inc.: Westport, Conn.,
    1975.
97. ANON. Nutr. Rev., 1978, 36, 346.
98. Gregory, J. F.; Kirk, J. R. J. Fd. Sci., 1978, 43, 1585.
99. Mason, J. B.; Gibson, N.; Kodicek, E. Brit. J. Nutr., 1973,
    30, 297.
100. Darby, W. J.; McNott, K. W.; Todhunter, E. N. Nutr. Rev.,
    1975, 33, 289.
101. ANON. Nutr. Rev., 1974, 32, 167.
102. O'Dell, B. L. Am. J. Clin. Nutr., 1969, 22, 1315.
103. ANON. "Recommended Dietary Allowances", 9th Ed.; Nat.
    Acad. of Sci.: Washington, D. C., 1980.
104. Fritz, J. C. Chem. Tech., 1976, 10, 644.
105. Oberleas, D.; Harland, B. F. "Zinc Metabolism: Current
    Aspects in Health and Disease"; Brewer, G. L.; Prasad,
    A. S., Eds; Alan R. Liss, Inc.: New York, 1977.
106. Erdman, J. W., Jr. Cereal Chem., 1980 (in press).
107. Martinez, W. H. "Evaluation of Proteins for Humans";
    Bodwell, C. E., Ed.; AVI Publ. Co. Inc; Westport, Conn.,
    1977.
108. O'Dell, B. L. "Soy Protein and Human Nutrition"; Wilcke,
    H. L.; Hopkins, D. T.; Waggle, D. H., Eds.; Academic
    Press: New York, 1979.
109. Erdman, J. W., Jr. JAOCS, 1979, 56, 736.
110. Erdman, J. W., Jr.; Forbes, R. M. Fd. Prod. Dev., 1977,
    11(10), 46.

111. Forbes, R. M.; Parker, H. M. <u>Nutr. Repts. Int.</u>, 1977, <u>15</u>, 681.
112. Erdman, J. W., Jr.; Weingartner, K. W.; Mustakas, G. C.; Schmutz, R. D.; Parker, H. M.; Forbes, R. M. <u>J. Fd. Sci.</u>, 1980, <u>45</u>, 1193.
113. Forbes, R. M.; Weingartner, K. E.; Parker, H. M.; Bell, R. R.; Erdman, J. W.; Jr. <u>J. Nutr.</u>, 1979, <u>109</u>, 1652.
114. Weingartner, K. E.; Erdman, J. W., Jr.; Parker, H. M.; Forbes, R. M. <u>Nutr. Repts. Int.</u>, 1979, <u>19</u>, 223.
115. Davies, N. T.; Olpin, S. E. <u>Brit. J. Nutr.</u>, 1979, <u>41</u>, 579.
116. Davies, N. T.; Hristic, V.; Flett, A. A. <u>Nutr. Repts. Int.</u>, 1977, <u>15</u>, 207.
117. Morris, E. R.; Ellis, R. <u>J. Nutr.</u>, 1980, <u>110</u>, 1037.
118. Franz, K. B. "Bioavailability of Zinc From Selected Cereals and Legumes"; Ph.D. Thesis; University of California: Berkeley, 1978.
119. Momcilovic, B.; Shah, B. G. <u>Nutr. Repts. Int.</u>, 1976, <u>13</u>, 135.
120. Oberleas, D. "Proceedings of the Western Hemisphere Nutrition Congress, IV"; Amer. Med. Assoc.: Chicago, IL, 1975.
121. Ranhotra, G. S.; Lee, C.; Gelroth, J. A. <u>Nutr. Repts. Int.</u>, 1978, <u>18</u>, 487.
122. Ranhotra, G. S.; Lee, C.; Gelroth, J. A. <u>Nutr. Repts. Int.</u>, 1979, <u>19</u>, 851.
123. Rotruck, J. T.; Luhrsen, K. R. <u>J. Agr. Fd. Chem.</u>, 1979, <u>27</u>, 27.
124. Steinke, F. H.; Hopkins, D. T. <u>J. Nutr.</u>, 1978, <u>108</u>, 481.
125. Okubo, K.; Waldrop, A. B.; Iacobucci, G. A.; Myers, D. V. <u>Cereal Chem.</u>, 1975, <u>52</u>, 263.
126. Ford, J. R.; Mustakas, G. C.; Schmutz, R. D. <u>JAOCS</u>, 1978, <u>55</u>, 371.
127. Omosaiye, O.; Cheryan, M. <u>Cereal Chem.</u>, 1979, <u>56</u>, 58.

Received September 5, 1980.

Enzyme Modification of Proteins

R. DIXON PHILLIPS and LARRY R. BEUCHAT

Department of Food Science, University of Georgia College of Agriculture
Experiment Station, Experiment, GA 30212

Functionality can be defined as the physio-chemical behavior
which proteins exhibit while interacting with other constituents
of multi-component food systems. The basis of this behavior is,
of course, the chemical nature of proteins: long chains of
covalently bonded amino acid residues, the side groups of which
may be charged, polar, or hydrophobic. These chains are coiled
and arranged into a variety of configurations and may be
associated with other chains through covalent or non-covalent
bonds. Such macro-systems assume shapes which vary from globular
to fiberous and exhibit a wide range of functional behavior, as
shown in Figure 1 and described elsewhere in this volume. See
also the extensive review by Kinsella (1).
    The topic of this chapter deals with the effect of proteolytic
enzyme action on functionality. As noted by Whitaker (2), protein
functionality can be altered by many different types of enzymes.
However, since the great majority of enzymes which modify protein
behavior in food systems are proteases, we will confine our
discussion to them.
    The use of proteases for modification of protein functionality
is an ancient art. Originally, the enzymes were either endogenous
to the food (e.g., aging of meat) or part of an added bio-system
(e.g. microbial cultures in cheeses) (3). With increased
knowledge of what enzymes are and how they work, their delibrate
isolation and addition to food systems became widely practiced.
The motivation for using proteases to alter protein functionality
in food is two-fold: (a) The need to convert food source proteins
to more palatable or useful forms (increasingly important with the
advent of "novel" proteins); and (b) the rather specific way in
which proteases accomplish this goal.
    Proteases exert their influences by catalyzing the cleavage
or, more rarely, the synthesis of peptide bonds (Figure 2). The
breaking of peptide bonds results in three major modifications:
(a) An increase in the number of polar groups ($-NH_4^+$, $-CO_2^-$),
and an increase in the hydrophilicity of the product; (b) a

0097–6156/81/0147–0275$06.00/0
© 1981 American Chemical Society

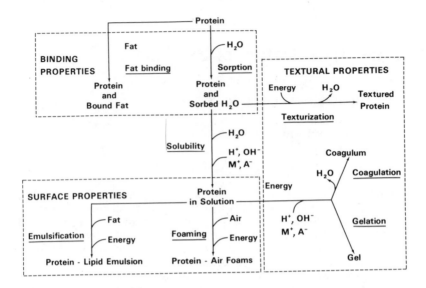

Figure 1. Aspects of protein functionality and their interrelationships

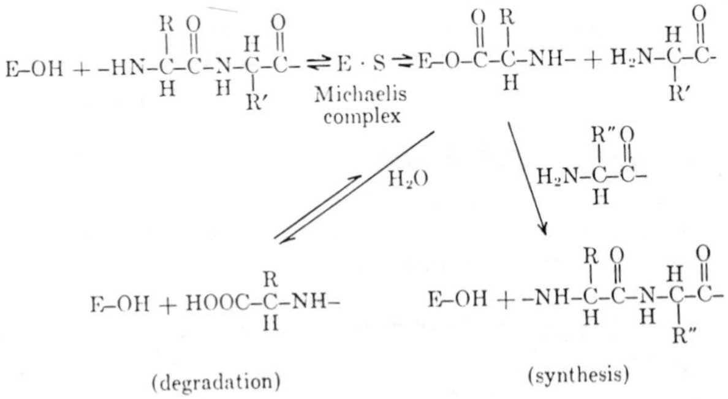

American Chemical Society

Figure 2. Formation of an acyl-enzyme intermediate by peptide bond scission and nucleophilic attack by an amine group to form a new peptide bond (transpeptidation), or by water to release the shortened chain (hydrolysis) (8)

decrease in molecular weight of the peptide chains, although not
necessarily of the total aggregates; (c) a possible alteration in
molecular configuration. Thus peptide bond breakage may be the
first step in a complex series of changes which alters
functionality, e.g., by dissociation of subunits or by opening a
compact globular structure to expose the hydrophobic interior to
an aqueous phase. The net synthesis of peptide bonds (as is
claimed in the plastein reaction) would be expected to decrease
the number of polar groups, and the hydrophilicity, to increase
molecular weight, and to affect configuration.

The foregoing description of the effects of proteases is
admittedly a simplified one when we consider the complexity of
food systems containing not only many constituents besides
protein, but also a great variety of different proteins. Likewise
the enzymes which modify functionality are usually multifuntional
mixtures, whether endogenous or exogenous. Nevertheless, this
description is helpful in understanding the effects of proteases
on functionality.

We have chosen to discuss enzyme modification of proteins in
terms of changes in various functional properties. Another
approach might have been to consider specific substrates for
protease action such as meat and milk, legumes and cereals, and
the novel sources of food protein such as leaves and
microorganisms (4). Alternatively, the proteases themselves
provide categories for discussion, among which are their source
(animals, plants, microorganisms), their type (serine-,
sulfhydryl-, and metalloenzymes), and their specificity (endo- and
exopeptidases, aromatic, aliphatic, or basic residue bond
specificity). See Yamamoto (3) for a review of proteolytic
enzymes important to functionality.

Plastein Reaction

The protease-catalyzed synthesis of peptide bonds is known as
the plastein reaction (5). Plastein itself is defined as the
product formed by this reaction which is insoluble in
trichloroacetic acid solutions (6). The plastein reaction has
been most extensively investigated by researchers in Japan (5, 6,
7, 8, 9). These scientists have reviewed various aspects of
plastein and the plastein reaction and its importance to protein
functionality and nutriture (6, 8, 10).

The conditions necessary for the plastein reaction have been
reviewed by Fujimaki et al. (8), and compared to those necessary
for proteolysis by Arai et al. (6). The substrate for the
synthetic reaction must consist of low molecular weight peptides,
preferably in the tetramer to hexamer range. These are usually
produced from proteins by protease action. A number of
proteolytic enzymes and protein substrates have been investigated
for producing plastein reaction substrates. The most often used
proteases are pepsin (9, 11), and papain (12, 13), but others

including acid, serine, and sulfhydryl proteases have been described (8). Substrate proteins have been derived from soy (14), casein and zein (6), gluten (7), milk whey (5, 15), microorganisms (5, 7, 9), leaves (16), and egg albumin (5, 11). Arai et al. (5) investigated the effects of substrate hydrophilicity and concluded that a combination of hydrophilic and hydrophobic peptides was optimal. They proposed a parameter, β/α, related to hydrophilicity and found that maximum yields of plastein are obtained when $\beta/\alpha \simeq 0.5$. Amino acids may also be incorporated along with peptides into plasteins provided they are present in the reaction mixture as esters. Aso et al. (12) studied the specificity of amino acid ester incorporation and found that the reaction velocity increased with the hydrophobicity of both the amino acid side chain and the alcohol moiety. In contrast to peptide bond cleavage but in keeping with the law of mass action, synthesis is promoted by high substrate concentration (20 - 40%, w/v). At <7.5% substrate concentration, no plastein formation is observed. Although proteases vary widely in their pH optima for bond cleavage, a relatively narrow range (4 - 7) is observed for the synthesis of peptide bonds for all proteases investigated.

It is generally agreed that new peptide bonds are formed during the plastein reaction. However, the reaction mechanism and the nature of resulting products remains controversial. The role of the protease is central to the reaction. The enzymes most thoroughly investigated for catalyzing plastein formation are α-chymotrypsin and papain (5, 12, 16). A number of other enzymes are discussed by Fujimaki et al. (8). Yamamoto (3) described the characteristics of protease-active sites. Amino acid residues are arranged in such a way as to accommodate the substrate and to destabilize the target bond (Figures 3 and 4). Serine proteases such as chymotrypsin and sulfhydryl proteases such as papain are similar in the way they accomplish this bond activation. The acyl-enzyme intermediates illustrated in Figures 3 and 4 are central to both lytic and synthetic reactions. The most straightforward mechanism for peptide bond formation would be reversal of proteolysis promoted by the very high substrate concentrations. The first step in such a mechanism would be formation of a peptidyl-enzyme intermediate with loss of water (the reverse of reaction b, Figure 3). Alternatively, the acyl-enzyme intermediate could be formed with scission of a peptide bond (Figure 2). The second step in both mechanisms would be nucleophilic attack by the amino group of a peptide or amino acid ester on the active acyl intermediate to form the new bond (Figures 2 and 4). The first of these mechanisms is condensation; the second, transpeptidation. There is evidence that both mechanisms are active in the plastein reaction (8, 15), although the later is thermodynamically favored (10).

The properties of plasteins are quite different from those of the starting peptide mixture. In addition to insolubility in

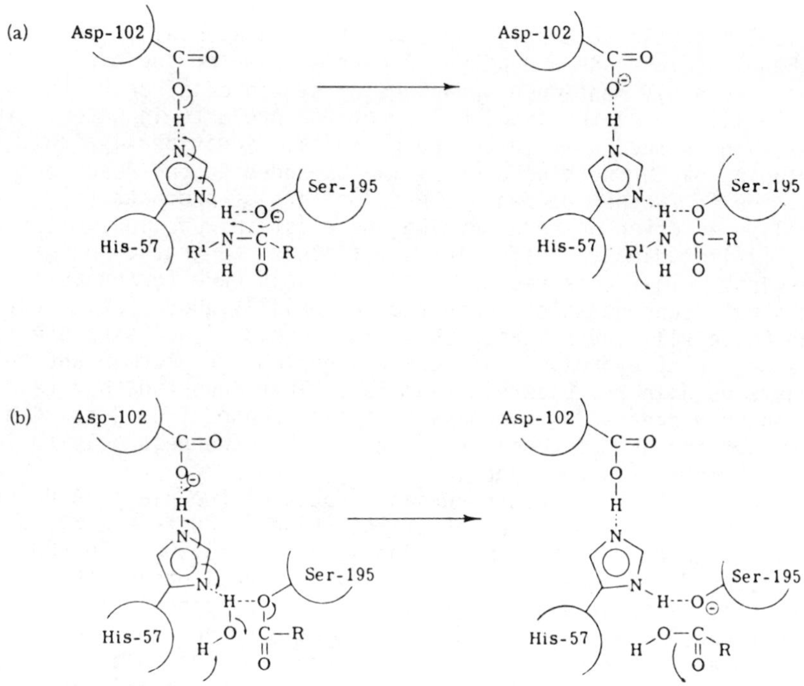

(a)   Asp-102 ⟩ C=O ... Ser-195 ... His-57 ...

(b)   Asp-102 ⟩ C=O ... Ser-195 ... His-57 ...

Nature

Figure 3.   Possible mechanism for (a) formation and (b) breakdown of acyl-enzyme (chymotrypsin) intermediate (3)

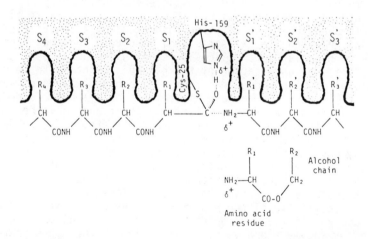

American Chemical Society

Figure 4.   Schematic of the active site of papain with peptide and amino acid ester in place (8)

trichloroacetic acid solutions, plasteins are usually insoluble in ethanol and acetone and often insoluble in water and buffers (6). The previously mentioned advantage of hydrophobic reactants is partially due to the insolubility of the products in water. The reaction is thus shifted toward products. Additionally, an organic solvent such as acetone may be added to the reaction mixture to further decrease the solubility of products (5). Gelling is often observed during the plastein reaction. Tsai et al. (17) studied the effect of substrate concentration on gel rheology. They observed that stronger gels were formed as substrate concentration increased, especially above 20%. Edwards and Shipe (11) investigated the relationship of gel strength to the degree of hydrolysis of pepsin-digested egg albumin and to the enzyme used in the plastein reaction. They reported that gel strength decreased as hydrolysis time increased from 4 to 24 hours and that the use of pepsin resulted in the strongest gels followed by α-chymotrypsin then papain.

Perhaps the most controversial aspect of the plastein reaction has been the molecular weight of resulting products. Early work (18) indicated very high molecular weights (250,000 - 500,000) as determined by ultra-centrifugal analysis. Arai et al. (6) found the upper limit of plastein formed from a soy protein hydrolyzate to be ∿30,000 as measured by gel exclusion chromatography. Hofsten and Lalasidis (15), however, reported that plasteins subjected to gel exclusion chromatography in 50% acetic acid showed no increase in molecular size over that of the reactants. Monti and Jost (19) reached the same conclusion based on gel chromatography in DMSO and on analysis of α-amino nitrogen in plasteins. Hofsten and Lalasidis (15) noted that hydrophobic peptides showed unusual elution behavior on sephadex gels in water or dilute buffers, providing a possible explanation for differences in their results compared to those of Arai et al. (6). Based on the solubility and chromatographic behavior of plasteins, Hofsten and Lalasidis (15) concluded that gel formation was due to nonpolar interactions of small hydrophobic peptides.

The original application of the plastein reaction to foods was the removal of bitter peptides from enzyme hydrolyzates of soy protein (6). Peptides containing aromatic or aliphatic side chain amino acids, especially in the terminal positions, have been identified with bitterness of enzyme hydrolyzates (8). Subjecting hydrolyzates to the plastein reaction eliminates this bitterness either by moving the hydrophobic residue to the interior of the peptide or by cleaving it during transpeptidation. Other functional improvements of proteins that may be accomplished by use of the plastein reaction have been reviewed by Fujimaki et al. (8). In addition to bitter peptides, unwanted impurities which contribute to off flavors, odors or colors may be eliminated from proteins by enzyme hydrolysis, purification of the hydrolyzate, and resynthesis by the plastein reaction (Figure 5). An improvement in the solubility profile of soy protein was

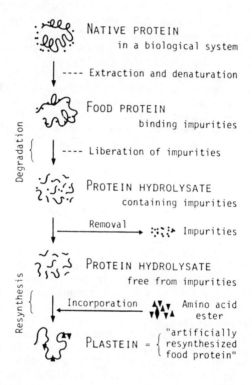

American Chemical Society

Figure 5. Scheme for removing impurities from protein substrate by hydrolysis, purification, and resynthesis via the plastein reaction (8)

demonstrated by protein-catalyzed incorporation of glutamic acid
into soy protein hydrolyzates (7). The production of food gels
could also be an important role for the plastein reaction (10).
Another major application of the plastein reaction is
nutritional improvement of proteins by the incorporation of
limiting amino acids (8). A plastein containing approximately 7%
methionine was produced from soy protein hydrolyzate and
L-methionine ethyl ester in the presence of papain. This material
was shown to be utilized as a source of methionine in the rat,
producing a PER of 3.38 when incorporated into soy protein diets
to give a methionine level of 2.74% of protein.

Similarly, lysine has been incorporated into gluten
hydrolyzate and lysine, threonine and tryptophan have been
individually incorporated into zein hydrolyzates. Lysine,
methionine, and tryptophan were incorporated simultaneously into
hydrolyzates of protein from photosynthetic origin. A very
interesting application of this procedure involved the preparation
of low-phenylalanine plasteins from a combination of fish protein
concentrate and soy protein isolate by a partial hydrolysis with
pepsin then pronase to liberate mainly phenylalamine, tyrosine,
and tryptophan, which were then removed on sephadex G-15. Desired
amounts of tyrosine and tryptophan were added back in the form of
ethyl esters and a plastein suitable for feeding to infants
afflicted with phenylketonuria was produced.

Eriksen and Fagerson (10), while citing the potential of the
plastein reaction to the food industry, listed several
constraints: (a) The possible problems of scale-up to industrial
sized operations; (b) the presence of residual enzymes; (c) the
possibility that toxic peptides could be produced; and (d) the
stigma attached to synthetic foods. Yamashita et al. (7)
mentioned economic constraints due to cost of enzymes.and
establishing safety, which must be determined through animal
feeding studies. Recent work has addressed some of these
constraints. Yamashita et al. (13) identified reaction conditions
that allowed the incorporation of methionine into soy protein in a
single step. This involved a reaction mixture that was 20% by
weight soy isolate, 1 nM in cysteine (enzyme activator), an
initial pH of 10, a papain to substrate ratio to 1:100, a
temperature of 37°C, and a reaction time 8 hours. When a 20:1
mixture of soy protein and methionine ethyl ester was incubated
under these conditions, 81.5 mole percent of the L-methionine was
incorporated.

Pallavicini et al. (16) utilized α-chymotrypsin immobilized on
chitin to catalyze plastein formation from leaf protein
hydrolyzates. When analyzed by gel exclusion chromatography, the
products were comparable to those produced by soluble enzymes.

Modification of Specific Functional Properties

Solubility. Puski (20) measured changes in various functional properties of soy protein isolate as related to the extent of treatment with a neutral protease from Aspergillus oryzae. Heat treatment of isolates decreased the solubility of nitrogen, but with increasing enzyme treatment less protein was rendered insoluble. Untreated controls did not show significant solubility at pH 4.5 in the presence of 0.03M CaCl$_2$, but proteins were increasingly soluble as treatment progressed. The author points out that increased acid solubility would be advantageous in the utilization of soy proteins in acidic foods, whereas calcium tolerance is important when calcium addition is needed for improved nutrition such as in imitation dairy products. Enzyme treatment to increase acid solubility (21) has also been described.

Exposure of proteins to heat has been shown to adversely affect peptization properties (22). The critical denaturation temperature for peanut protein in meal is above 118°C (dry heat) and above 80°C at 100% relative humidity. Although heat treatments routinely employed in a majority of peanut oil mills are insufficiently controlled to prevent denaturation of meal protein, the authors suggest that temperature and moisture control during processing can be maintained at levels sufficient to achieve oil extraction without drastic denaturation of protein.

Better and Davidsohn (23) reported that the susceptibility of proteins in oilseed presscake and solvent-extracted meal was strongly affected by heat treatment. Working with peanut and coconut presscake, they demonstrated that the stronger the heat treatment, the higher are the concentrations of pepsin required to make peptization more complete. This is an important fact to bear in mind when oilseed press or extraction residues are being used as raw materials for production of industrial protein.

Nitrogen solubility of defatted peanut flour as affected by hydrolysis with pepsin, bromelain and trypsin was investigated by Beuchat et al. (24). Profiles of enzymatically hydrolyzed flour in water were markedly different from their respective controls. Solubilities for pepsin, bromelain-, and trypsin-treated samples were 64, 46, and 30% of total nitrogen, respectively, at pH 4.0, compared to less than 15% for nontreated flours. Although hydrolyzed samples had higher nitrogen solubilities than their respective pH-adjusted controls in the alkaline pH range, they were less soluble than samples that had received no heat or pH adjustment. The influence of change in ionic strength of peanut flour slurry resulting from the addition of HCl and NaOH during pH adjustment was cited as a possible reason for alterations in nitrogen solubilities at alkaline pH values.

Beuchat et al. (24) also examined the nitrogen solubility profiles for enzymatically hydrolyzed and control peanut flour samples in 0.03M Ca^{2+} (as CaCl$_2$). Solubilities of controls between pH 2.0 and 5.0 in 0.03M Ca^{2+} solutions were similar to those noted for water; however, very little increase in nitrogen solubility of controls was noted in the pH 5.0 to 11.0 range.

Similar data were reported by Rhee et al. ($\underline{25}$). They noted that at pH 2.0 to 3.0, peanut protein extractability was enhanced in 0.01 to 0.10M $CaCl_2$ but suppressed in solutions containing 0.25M or higher $CaCl_2$. Treatment of peanut flour with pepsin, bromelain and trypsin greatly increased the nitrogen solubility at pH 2.0 to 11.0 in 0.03M Ca^{2+} ($\underline{24}$). Pepsin treatment resulted in the greatest increase. Lowest solubilities in the profiles were at pH 4.0 to 5.0, where values of 81, 57, and 38% were measured for pepsin-, bromelain-, and trypsin-treated flours, respectively.

A concentration of 0.03M Ca^{2+} has been prescribed as a minimum in the formulation of imitation milk. Evidence from the study reported by Beuchat et al. ($\underline{24}$) suggests that enzymatic hydrolysis of peanut flour modifies protein to the extent that it is highly soluble in 0.03M Ca^{2+} at a pH range normally associated with fluid milk. Further studies are required to assess the effect of enzyme-induced proteolysis on organoleptic properties of hydrolyzed peanut protein solutions.

Changes in soluble proteins of peanut flour caused by proteolytic enzyme digestion as detected by gel electrophoresis ($\underline{24}$) resemble those resulting from fermentation of peanut meal with fungi ($\underline{26}$) and growth of fungi on viable peanut kernels ($\underline{27}$). "Standard" peanut protein electrophoretic patterns are distinctly modified as a result of treatment with commercial proteases or fungal growth. Biochemical transformations in proteins include decomposition of large molecular weight globulins such as arachin to smaller components, followed by rapid quantitative and qualitative decreases in these latter constituents as fungal enzymatic activity progresses. Thus, nitrogen solubility changes dramatically.

The use of live fungal cultures rather than enzymes extracted therefrom to modify the functionality of defatted peanut flour was investigated by Quinn and Beuchat ($\underline{28}$). Fungi not known to produce mycotoxins were studied: Aspergillus elegans, A. oryzae, Mucor hiemalis, Neurospora sitophila, and Rhizopus oligosporus. Each of the fungi increased the nitrogen solubility of peanut flour in the pH range of 3.0 to 6.0 compared to the control. In particular, M. hiemalis increased the solubility at pH 4.0 to 5.0 from less than 5% in a nontreated control to about 34% in the freeze-dried ferment. Although ferments were demonstrated to be nontoxic, there may be special regulatory problems associated with incorporation of new microbially-derived or modified ingredients into food products for the purpose of functionality.

Sekul et al. (29) studied the nitrogen solubility properties of enzyme-hydrolyzed peanut proteins. A deionized water dispersion of peanut flour (1:10, w/v) was treated with papain (0.5% total volume) at 45°C for 15 min. Solubility was tested over a range of pH 1 to 9. In general, papain treatment improved solubility at all levels examined except pH 2 and 8 (Figure 6). The authors suggest that hydrolysis of peanut proteins in aqueous slurries of defatted flour by papain should be of commercial interest. For beverage application where high protein solubility is desired (milk-type drinks, acid pH fruit-flavored beverages, dry soup, sauce, or gravy mixes), partially hydrolyzed peanut proteins seem to have an advantage over unhydrolyzed proteins. Also, papain is one of the least expensive FDA-approved vegetable enzymes; there are economic or FDA health-related restrictions on the addition of animal or microbial enzymes to food formulations; and studies have shown that papain is free of peptidase activity (30).

Fontaine et al. (31) presented data comparing the solubility behavior of proteins of peanut and cottonseed meals, proteins of corresponding dialyzed meals, and isolated proteins. While the shapes of the pH/solubility curves for cottonseed and peanut meals differed, the response of proteins to the removal of dialyzable meal constituents was similar. Data indicated the presence of natural materials in both meals which decreased the solubility of meal nitrogen at certain acid pH values but exerted no effect at alkaline pH values. Thus procedures for solubilizing proteins by treatment with proteolytic enzymes should also be designed with consideration of the influence of non-protein constituents.

The relative activities of commercially available proteolytic enzymes and hemicellulase were tested on cottonseed cake by Arzu et al. (32). These included ficin concentrate, three bacterial proteinases, papain concentrate, bromelain concentrate, pepsin, trypsin, acid fungal protease, and fungal hemicellulase. The most active enzymes were two bacterial proteinases and the ones derived from higher plants (bromelain, papain, and ficin), which solubilized about 40% of the initial insoluble protein. Activity shown by the hemicellulase was the lowest, barely facilitating the solubilization of nitrogenous compounds, contrary to what has been reported by Hang et al. (33, 34) on mung and pea beans employing cellulase, where a large solubilization was noticed. Gossypol markedly inhibited the activity of bacterial proteases.

Sreekantiah et al. (35) evaluated defatted sesame and peanut meals, and four kinds of beans - chickpea, green gram, black gram, and field bean - for protein extraction characteristics after treatment with enzymes. Three commercial proteolytic enzymes derived from Aspergillus species and one from Trametes sanguina were tested. The protease from T. sanguina was best for hydrolyzing sesame meal and chickpea. In the case of sesame meal, the soluble protein increased fom 9.48 to 64.12% while the increase in amino nitrogen was from 0.6 to 9.2%. Appreciable increases, both in soluble and amino nitrogen, were observed in treated bengal gram. Results showed that uncooked substrates were not attacked by enzymes. Adjustment of pH to 3.0 by dilute acid after heat treatment was congenial for the hydrolysis of protein in sesame meal and chickpea. The amino acid composition of extracts of enzyme-hydrolyzed proteins may differ somewhat from that of untreated materials. In the case of sesame meal, there

was an increase in the percentage of all essential amino acids
except methionine, when compared to the defatted meal. In
chickpea, even though the percentage of protein in the
freeze-dried extract was higher, there was a reduction in lysine,
histidine, arginine and leucine.

Hermansson et al. (36) used pepsin and papain to solubilize
rapeseed protein concentrate. Papain had a lower solubilizing
effect than did pepsin. However, the fact that pepsin has an
optimum pH for activity at about 1.6, far below the pH range of
most foods, made it possible to study the effects of controlled
hydrolysis. At pH 7.0, all hydrolysates were more soluble than
the original rapeseed protein concentrate.

The use of enzymes for solubilization of seed and leaf
proteins has been studied as a means of overcoming difficulties
presented by the varying condition of seed and leaf material
available for processing (37). Ultrasonic energy was reported to
increase the efficiency of enzyme solubilization procedures. The
effects of various conditions of hydrolysis of cottonseed and
alfalfa meal protein with trypsin were defined.

Several proteolytic enzymes have been shown to enhance the
solubility of fish protein concentrate (38). Product inhibition
and self destruction of enzymes occurred, so that rates of
hydrolysis decreased with time.

The elimination or inactivation of enzymes used to treat
proteins is a critical problem once the desired modification in
functionality is achieved. In many instances, product inhibition
or self destruction does not occur as noted above for fish protein
concentrate. As stated by Puski (20), if heat inactivation is
used, the proteins may be denatured and revert to insoluble
forms. Washing out the enzyme at its isoelectric point would also
remove a portion of the protein which is solubilized by the
enzyme. Inactivation of enzymes by chemical means may also cause
significant changes in the protein. Thus, while desired
functional modifications of food ingredients may be obtained
through enzyme treatment, the problem of latent enzyme activity in
food formulations must be addressed.

Emulsifying Capacity and Stability. Zakaria and McFetters
(39) studied the effects of pepsin hydrolysis of heated soy
protein isolate. Treatment was carried out between pH 1.0 and
4.0. At short hydrolysis times, there was a rapid increase in the
free amino groups in the protein and a corresponding increase in
emulsification activity (expressed as the volume percentage of the
emulsified layer). The emulsification activity declined as the
incubation time was increased (Figure 7). However, only small
increases in free amino groups occurred during the period of
decreased activity. Good emulsion stability properties were
observed in hydrolysates with high emulsion activity. The authors
suggest that either hydrolysis of a few key soy protein peptide
bonds results in relatively large changes in emulsification

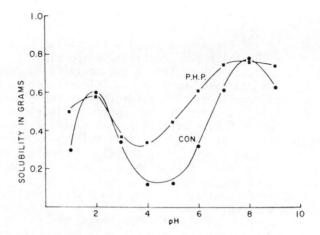

Figure 6.   Nitrogen solubility (Kjeldahl) of peanut proteins: P.H.P. = partial hydrolysis by 5% papain; CON. = control (no papain) (28).

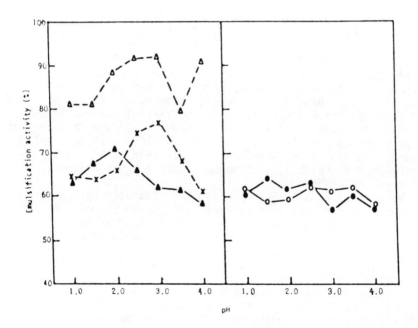

Libbensmittel–Wissenschaft & Technologie

Figure 7.   Emulsifying activity (expressed as volume percentage of the emulsified layer) of pepsin hydrolysates of soy protein as a function of pH and hydrolysis time. No hydrolysis (▲); 2 h (△); 8 h (×); 17 h (●); 24 h (○) (39).

properties or that changes other than proteolysis occur which affect functionality.

Adler-Nissen and Olsen (40) studied the influence of peptide chain length on the taste and functional properties of enzymatically modified soy protein. The emulsifying capacity of modified proteins could be improved significantly compared to unmodified control samples by controlling the extent of hydrolysis.

The emulsifying capacity of soy protein isolate can be increased by treatment with neutral fungal protease (20); however, enzyme treatment decreased emulsifying stability. It was theorized that enzyme digestion of proteins increases the number of peptide molecules available at the oil-water interface, and thus, a larger area may be "covered", resulting in the emulsification of more oil. However, since these peptides are smaller and less globular compared to those in untreated soy isolate, they may form a thinner layer around the oil droplets, thus, resulting in a less stable emulsion. Enzyme treatment may have also exposed buried hydrophobic groups which resulted in improved hydrophilic-lipophilic balance for better emulsification. Burnett and Gunther (41) and Gunther (42) have patented procedures for the commercial production of a whipping protein by partial hydrolysis of soy protein.

The emulsifying capacity and the viscosity of emulsions of peanut proteins partially hydrolyzed with papain are reported to be generally lower than for unmodified flour (29). Products from papain hydrolysis of peanut protein differ substantially from those described by Beuchat et al. (24), who used bromelain, pepsin, and trypsin. These researchers reported that emulsion capacities of peanut flour were higher in water than in 0.5M NaCl and that heating did not result in substantial changes in the capacity of flour to emulsify oil. However, enzymatic digestion of proteins completely destroyed the emulsifying capacity of the flour. Apparently hydrolysis altered protein surface activity strengths and the ability of peanut protein to stabilize oil-in-water emulsions.

Hermansson et al. (36) examined the stability ratings for protease-treated emulsions of rapeseed protein concentrate at pH 5.5 and 7.0. The use of carboxymethylcellulose in conjunction with pepsin hydrolysis gave approximate ten-fold increases in stability ratings for emulsions.

Kuehler and Stine (43) studied the functional properties of whey protein with respect to emulsifying capacity as affected by treatment with three proteolytic enzymes. Two microbial proteases and pepsin were examined. The emulsion capacity decreased as proteolysis continued, suggesting that there is an optimum mean molecular size of the whey proteins contributing to emulsification.

Considerable effort has been devoted to the improvement of functional properties of fish protein concentrate. Spinelli et al. (44) conducted a study to determine the feasibility of modifying myofibrillar fish proteins by partially hydrolyzing them with proteolytic enzymes. Increases in both emulsion stability and capacity (g oil/g protein) were achieved by partial hydrolysis. The emulsion capacity, as well as solubility, of spray-dried enzyme-modified myofibrillar protein were about 5% lower than those of freeze-dried samples. The authors concluded that fish protein isolates possessing valuable functional properties can be prepared from myofibrillar proteins that are partially hydrolyzed with a proteolytic enzyme. Caution is advised, however, regarding the long-term storage stability of such products due to chemical alterations of residual lipids.

Foaming Capacity and Stability. Pepsin digestion of soy protein has been proposed as a method for making a whipping protein for egg albumen replacement (42, 45) and for extenders for albumen in bakery and confectionery formulations (46). Puski (20), on the other hand, reported that although treatment of soy protein isolate with fungal protease increased foam volume, stability was reduced to zero. The lack of stability was attributed in part to heat treatment which may have denatured the large protein components sufficiently so that they could not act as a stabilizing component.

Limited digestion of globular soy proteins with rennin affords a modified protein preparation which retains a high molecular weight (47). Whipping quality, measured by foam volume and stability, was superior in comparison with native proteins. The limited rennin proteolysis of soy was identified as a key factor in functionality, since this modification conferred improved solubility.

Partial hydrolysis of peanut proteins with papain significantly increases both foaming capacity and foam volume (29). When the pH was first adjusted to 4, then back to 8.2, foaming capacity increased three-fold. While the stability of foam at acid pH (6.3/6.7) was low (less than 30 min), stability at pH 8.2 was long lasting, suggesting greater potential for partially hydrolyzed peanut proteins in nonacidic pH foods. The authors suggested that papain-modified peanut protein flour should find use in products such as frozen desserts, soft mix ice creams, dessert and pie toppings.

Pepsin and papain hydrolysates of rapeseed protein concentrate increased foam volumes and decreased drainage compared to the untreated control (36). Foaming properties could be further enhanced by adding a stabilizer such as carboxymethylcellulose.

Foaming or whippability characteristics of whey protein as affected by treatment with three proteolytic enzymes were evaluated by Kuehler and Stine (43). The specific volume of foams increased initally as a result of treatment, then decreased with time at a more rapid rate compared to nontreated whey (Figure 8). A limited amount of hydrolysis appears to be desirable to increase foam volume but foam stability is greatly decreased as a result of such hydrolysis. The authors suggest that this is probably due to

increasing the polypeptide content initially which allows more air
to be incorporated. However, the polypeptides do not have the
strength required to maintain a stable foam and further hydrolysis
likely results in peptides which lack any capacity to stabilize
the air cells of the foam. Limited proteolysis may be
advantageous for utilizing whey proteins in foams since specific
volume was increased by as much as 25%. The decrease in stability
which results from limited hydrolysis can be retarded by adding
stabilizers such as carboxymethylcellulose.

Proteolytic enzymes derived from Aspergillus oryzae and
Streptomyces griseus enhance the foaming capacity of frozen whole
egg products (48). More recently Grunden et al. (49) examined the
effects of several crude proteolytic enzymes on functional
properties of egg albumen, as measured by angel food cake volume
and foam volume and stability. Papain, bromelain, trypsin, ficin,
and a fungal protease were evaluated. Volumes of all cakes
containing enzyme-treated albumen were comparable to or better
than the control. Differences were noted in the texture of angel
food cakes made from control albumen and those made from
enzyme-treated albumen. Unlike the control cakes, those
containing hydrolyzed albumen had a coarse and gummy texture. Off
odors and flavors were detected in cakes made from albumen that
had been treated with bromelin or fungal protease.

Treatment of albumen with trypsin, bromelain and fungal
protease produced significantly greater volumes of foam compared
to stored control albumen (49). However, all enzyme treatments
had inferior foam stability when compared to controls. Both the
rate and amount of foam collapse was greater in enzyme-treated
samples. The fresh control produced the most stable foam.

Changes in aeration properties of bromelain-modified
succinylated fish proteins were studied by Groninger and Miller
(50). The effects of whipping time, protein concentration, pH and
additives (salt, sugar, vanialla flavoring and fat) on foam volume
and stability were determined. Volume and stability were
increased as a result of modification. In comparison to egg white
and soy protein, enzyme-hydrolyzed, succinylated fish protein had
a lower foam volume but was more stable. Hydrolyzed, succinylated
protein formed foams over the pH range of 3.0 to 9.0; foams
produced at a pH below 7.0 showed lower stability. Also, there
was not a significant decrease in the whipping properties of the
hydrolyzed, succinylated fish protein in the isoelectric range of
approximately pH 4.5. This is in contrast to unhydrolyzed soy
protein, for example, which has a pronounced decrease in foam
volume and stability in the isoelectric range (51).

Additives had substantial effects on the aeration properties
of bromelain-modified succinylated fish protein (50). Foam volume
was increased with up to 2% sodium chloride in the system; however
there was a decrease in foam stability when greater than 0.3% salt
was used. Sucrose, at concentrations up to 50%, increased foam
stability. When fat was added to treated fish protein dispersions

previous to whipping, foam formation was inhibited. When fat was added to a protein foam after it was whipped to a maximum volume, the foam flattened.

Water Uptake and Retention. The water binding capacity of soy protein isolate can be increased by treatment with neutral fungal protease (20). Since the number of free amino and carboxyl groups increases as a result of digestion and because moisture uptake by proteins is proportional to the number of ionic groups present (52), it is not surprising that moisture uptake is increased by enzyme treatment.

The effects of fungal fermentation on the moisture adsorption and retention properties of defatted peanut flour have been reported (28). At 8 and 21°C, little difference was noted between the moisture contents of freeze-dried ferments and untreated samples equilibrated at relative humidities ranging from 14 to 75%. However, marked changes were noted above 75% equilibrium relative humidity (ERH), where the control samples did not adsorb as much moisture as did the ferments. These differences were attributed to an increased ratio of exposed hydrophilic to hydrophobic groups resulting from fermentation.

Changes in the capacity of defatted peanut flour to adsorb moisture as a result of treatment with protease were investigated by Beuchat et al. (24). Hydrolysis with pepsin, bromelain, and trypsin caused the peanut flour to adsorb more water than nontreated controls at specific ERH values. Increased water adsorbing capacities of enzyme-treated peanut proteins are probably related to increased numbers of polar sites, such as carboxyl and amino groups, which appear on proteins as a result of hydrolysis. Peptization and permanent configurational changes may occur during heat treatment or exposure to acidic and alkaline pH conditions, influencing the capacity of seed flours to adsorb moisture. Better and Davidsohn (23) noted that heat and pH may alter the effectiveness of using pepsin to solubilize proteins in peanut meal.

From a practical viewpoint, the increased water-adsorbing capacity of enzyme-treated flour at specific ERH values may have important implications in the formulation of intermediate-moisture foods. At an ERH of 60%, a level regarded as minimal for the growth of microorganisms in foodstuffs, Beuchat et al. (24) reported that the equilibrium moisture content of nontreated peanut flour was 10% compared to 14% for flour that had been treated with pepsin (Figure 9). Conceivably, a food product could be developed using pepsin-digested flour which would be safe from the standpoint of not supporting microbial growth and yet contain significantly more water than a product formulated using nontreated flour. Such a product might not exhibit the characteristic mouth-drying sensation often associated with vegetable proteins.

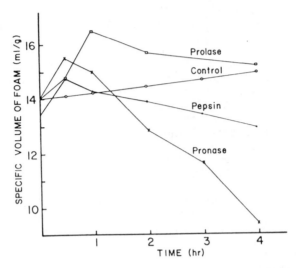

Figure 8.   Effect of enzymatic hydrolysis on specific volume of foam obtained by whipping a heated whey protein sol (4% w/w, 85°C, 6 min whipping) (43)

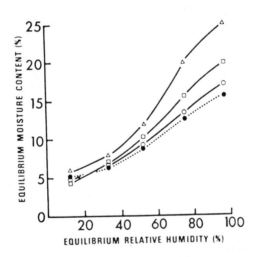

Figure 9.   Moisture adsorption isotherms of enzyme-treated and nontreated peanut flour: no pH-heat treatment (● – – – ●); 50-min treatment at 50°C followed by 10 min at 90°C (——); pepsin, pH 2.0 (△); bromelain, pH 4.5 (□); trypsin, pH 7.6 during heat treatment. All samples were adjusted to pH 6.9 prior to freeze-drying and analytical examination (24).

Application to Food Systems

Fungal enzymes have been used for hundreds of year to prepare and modify foodstuffs. However, modern industrial enzyme technology probably started with Takamine (53) and his work with A. oryzae. Today many industrial enzymes used to modify functional properties of foods and food ingredients are of fungal origin (54).

Microbial proteolytic enzymes have been used for decades to improve functional characteristics of aged wheat flours. The role of proteases in relation to production of bread was reviewed in 1946 (55). Improvement in quality has been evidenced by changes in handling quality of doughs, improved elasticity and texture of the gluten washed from such doughs, and substantial increases in loaf volume of bread (56). Swanson and Andres (57) reported that the amount of gluten that could be recovered from doughs treated with papain decreased with increased enzyme concentration and the rest period of the doughs was lengthed. Oka et al. (58) studied the action of pepsin on glutenin and reported that the cleavage of only a few peptide bonds rapidly produced large molecular polypeptides.

Verma and McCalla (56) studied the action of pepsin, papain and a commercial fungal protease on wheat gluten. All enzymes acted effectively on dispersed gluten; however, the action of different enzymes produced different types of digestion products. Depending upon desired handling characteristics of bread doughs prepared from treated wheat flour, various types of protease treatments can be selected.

Proteases are used in the baking industry to improve handling properties of doughs, elasticity and texture of gluten, and loaf volume (59). Fungal proteinases are widely used and have assumed a more important role than the amylases due to the general low level of naturally occurring proteinases in flour, to strict control of mixing times, and the desirability of doughs with optimum handling properties. Bacterial proteinases are used in general food production and in the production of baked goods such as crackers, cookies, and many snack foods. Bromelain and papain are approved additives but have not been widely used for the production of white bread.

Beuchat (60) investigated the performance of enzyme-hydrolyzed defatted peanut flour in a cookie formula. Flour slurries were treated with pepsin at pH 2.0, bromelain at pH 4.5, and trypsin at pH 7.6. After readjustment to pH 6.9, materials were freeze-dried, pulverized (60-mesh), and then substituted for wheat flour at 5, 15, and 25%. Adjustment of peanut flour to pH 2.0, as well as treatment with pepsin at this pH, greatly improved the handling characteristics of dough in which these flours were incorporated. Use of peanut flours treated at pH 4.5, with or without bromelain, and at pH 7.6, with or without trypsin, improved handling properties of cookie dough. These doughs did not tend to crumble

and compared favorably in texture to that of 100% wheat flour
control.

Enzyme hydrolysis of peanut flour also altered the physical
characteristics of baked cookies (60). With the exception of the
bromelain hydrolysate, the use of peanut flour in cookies resulted
in increased specific volume when compared to the 100% wheat flour
control. Untreated peanut flour substitution reduced the diameter
and increased the height of cookies; however, treatment with
proteolytic enzymes reversed the behavior. As evidenced by
substantial increases in spread ratios, the diameter of cookies
containing treated flours increased proportionately more than did
the height. These data promote the feasibility of decreasing or
increasing the spread of cookies through the addition of various
amounts of untreated or enzyme-treated peanut flour.

The functionality of bromelain-hydrolyzed succinylated fish
protein has been tested in a dessert topping, a souffle, and both
chilled and frozen desserts (50). Taste panel evaluations
revealed that no fishlike odors or flavors were detected. The
compatibility of enzyme-treated fish protein in these diverse food
systems points up the potential value of such a product to the
food processing industry.

Kinsella (1) reviewed the principal categories of functional
properties of proteins in foods and outlined various factors
affecting them from the point of view of the food chemist and
technologist.

Enzymatic modification of proteins applicable to foods is
reviewed by Whitaker (2). Described briefly are present uses of
proteolytic enzymes for modifying proteins through partial
hydrolysis. Major emphasis is placed on those enzymes which bring
about aggregation of proteins, cross-link formation, and side
chain modification through post-translational changes in the
polypeptide chain.

Richardson (4) reviewed the changes in proteins following
action of enzymes. Various exogenous and endogenous enzymes which
alter functional properties are discussed. Selected food systems,
e.g., meat, milk, and dough proteins, are emphasized to relate the
extent to which proleolytic enzymes play a role in modifying and
influencing functional properties of foods. The need for basic
research directed toward determining the importance of structural
features involved in various types of protein functionality is
stressed. It is also pointed out that empirical functional tests
on protein should be correlated with performance of the protein in
given food products.

Conclusions

The expanded use of enzymes to modify protein functional
properties has great promise for the food industry. Major
advantages of using proteases compared to other agents include
their specificity, their effectiveness at low concentrations and

under mild conditions, and their general safety, thus eliminating the necessity for removing them from finished products in most instances.

Existing uses of proteases in foods have been discussed in the foregoing section. Expanding such applications in the future depends upon our ability to control both the processes themselves and their costs. The development of continuous reactors utilizing free or immobilized enzymes will address each of these constraints. Furthermore, our understanding of the chemical basis for the various functional properties of proteins must be expanded (4). That is, we must learn how amino acid content and molecular configuration of food proteins are related to their functional properties. This goal is made more difficult by the fact that secondary, tertiary, and quarternary structures of proteins are likely to be quite different when exerting functional effects in food systems as compared to structures of the same proteins in dilute solutions and in their native states. The way in which specific actions of proteases affect protein structure must also be studied so that correlations with changes in functional properties can be made (61).

Literature Cited

1. Kinsella, J. E. Crit. Rev. Food Sci. Nutr., 1976, 7, 219.

2. Whitaker, J. R. In "Food Proteins. Improvement Through Chemical and Enzymatic Modification;" Gould, R. F., Ed. Am. Chem. Soc.: Washington, D.C., 1977; p. 95.

3. Yamamoto, A. In "Enzymes in Food Processing;" Reed, G., Ed., Academic Press: N.Y., 1975; p. 123.

4. Richardson, T. In "Food Proteins. Improvement Through Chemical and Enzymatic Modification;" Gould, R. F., Ed. Am. Chem. Soc.: Washington, D.C., 1977; p. 185.

5. Arai, S.; Yamashita, M.; Aso, K.; Fujimaki, M. J. Food Sci., 1975, 40, 342.

6. Arai, S.; Yamashita, M.; Fujimaki, M. Cereal Foods World, 1975, 20, 107.

7. Yamashita, M.; Arai, S.; Fujimaki, M. J. Agric. Food Chem., 1976, 24, 1100.

8. Fujimaki, M.; Arai, S.; Yamashita, M. In "Food Proteins: Improvement Through Chemical and Enzymatic Modification;" Gould, R. F., Ed. Amer. Chem. Soc.: Washington, D.C., 1977; p. 156.

9. Fujimaki, M., Kato, H.; Arai, S.; Yamashita, M. J. Appl. Bacteriol., 1971, 34, 119.

10. Eriksen, S.; Fagerson, I. S. J. Food Sci., 1976, 41, 490.

11. Edwards, J. H.; Shipe, W. F. J. Food Sci. 1978, 43, 1215.

12. Aso, K.; Yamashita, M., Arai, S.; Suzuki, J.; Fujimaki, M. J. Agric. Food Chem., 1977, 25, 1138.

13. Yamashita, M.; Arai, S.; Imaizummi, Y.; Amano Y.; Fujimaki, M. J. Agric. Food Chem., 1979. 27, 52.

14. Arai, S.; Yamashita, M.; Fujimaki, M. Cereal Chem., 1974, 51, 143.

15. Hofsten, B. V.; Lalasidis, G. J. Agric. Food Chem., 1976, 24, 460.

16. Pallavicini, C.; Finley, J. W.; Stanley, W. L.; Watters, G. G. J. Sci. Food Agric., 1980, 31, 273.

17. Tsai, S.; Yamashita, M.; Arai, S.; Fujimaki, M. Agric. Biol. Chem., 1972, 36, 1945.

18. Tauber, H. J. Am. Chem. Soc., 1951, 73, 1288.

19. Monti, J. C.; Jost, R. J. Agric. Food Chem., 1979, 27, 1281.

20. Puski, G. Cereal Chem., 1975, 52, 655.

21. Pour-El, A.; Swenson, T. S. United States Patent No. 3,713,843, 1973.

22. Fontaine, T. D.; Samuels, C.; Irving, G. W. Ind. Engineering Chem. 1944, 36, 625.

23. Better, E.; Davidsohn, A. Oleagineux, 1958, 13, 79.

24. Beuchat, L. R.; Cherry, J. P.; Quinn, M. R. J. Agric. Food Chem., 1975, 23, 616.

25. Rhee, K. C.; Cater, C. M.; Mattil, K. F. J. Food Sci., 1972, 37, 90.

26. Beuchat, L. R.; Young, C. T.; Cherry, J. P. Can. Inst. Food Sci. Technol. J., 1975, 8, 40.

27. Cherry, J. P.; Mayne, R. Y.; Ory, R. L. Physiol. Plant Pathol., 1974, 4, 425.

28. Quinn, M. R.; Beuchat, L. R. J. Food Sci., 1975, 40, 475.

29. Sekul, A. A.; Vinnett, C. H.; Ory, R. L. J. Agric. Food Chem., 1978, 26, 855.

30. Sekul, A. A.; Ory, R. L. J. Am. Oil Chem. Soc., 1977, 54, 32.

31. Fontaine, T. D.; Irving, G. W.; Markley, K. S. Ind. Engineering Chem., 1946, 38, 658.

32. Arzu, A.; Mayorga, H.; Gonzalez, J.; Rolz, C. J. Agric. Food Chem., 1972, 20, 805.

33. Hang, Y. D.; Wilkens, W. F.; Hill, A. S.; Steinkraus, K. H.; Hackler, L. R. J. Agric. Food Chem., 1970, 18, 9.

34. Hang, Y. D.; Wilkens, W. F.; Hill, A. S.; Steinkraus, K. H.; Hackler, L. R. J. Agric. Food Chem., 1970, 18, 1063.

35. Sreekantiah, K. R.; Ebine, H.; Ohta, T.; Nakano, M. Food Technol., 1969, 23, 1055.

36. Hermansson, A.-M.; Olsson, D.; Holmberg, B. Lebens.-Wiss. u. -Technol., 1974, 7, 176.

37. Childs, E. A.; Forte, J. L.; Ku, Y. In "Enzymes in Food and Beverage Processing;" Ory, R. L.; St. Angelo, A. J., Ed. Am. Chem. Soc.: Washington, D.C., 1977; p. 304.

38. Cheftel, C.; Ahearn, M.; Wang, D.; Tannenbaum, S. R. J. Agric. Food Chem., 1971, 19, 155.

39. Zakaria, F.; McFeeters, R. F. Lebens.-Wiss. u.-Technol., 1978, 11, 42.

40. Alder-Nissen, J.; Olsen, H.S. In "Functionality and Protein Structure;" Pour-El, A., Ed. Am. Chem. Soc.: Washington, D.C., 1979; p. 125.

41. Burnett, R. S.; Gunther, J. K. U.S. Patent No. 2,489,173, 1947.

42. Gunther, R. C. Canadian Patent No. 905,742, 1972.

43. Kuehler, C. A.; Stine, C. M. J. Food Sci., 1974, 39, 379.

44. Spinelli, J.; Koury, B.; Miller, R. J. Food Sci., 1972, 37, 604.

45. Turner, J. R. United States Patent No. 2,489,208, 1969.

46. Fox, K. K. In "Casein and Whey Products in By-Products from
    Milk;" Webb, B. H., Ed. AVI Publ. Co.: Westport, CT,
    1970, p. 331.

47. Lewis, B. A.; Chen, J. H. In "Functionality and Protein
    Structure;" Pour-El, A., Ed. Am. Chem. Soc.:
    Washington, D.C., 1979, p. 27.

48. Kewpie, K. K. Japanese Patent No. 40,260/70, 1970.

49. Grunden, L. P.; Vadehra, D.V.; Baker, R. C. J. Food Sci.
    1974, 39, 841.

50. Groninger, H. S.; Miller, R. J. Food Sci., 1975, 40, 327.

51. Eldridge, A. C.; Hall, P. K.; Wolf, W. J. Food Technol.,
    1963, 17, 1592.

52. Mellon, E. F.; Korn, A. H.; Hoover, S. R. J. Am. Chem. Soc.,
    1947, 69, 827.

53. Takamine, J. United States Patent Nos. 525,820 and 525,823,
    1894.

54. Bothast, R. J.; Smiley, K. L. In "Food and Beverage
    Mycology;" Beuchat, L. R., Ed. AVI Publ. Co.: Westport,
    CT, 1978; p. 368.

55. Hildebrand, F. C. In "Enzymes and Their Role in Wheat
    Technology;" Anderson, J. A., Ed. Interscience: New
    York, 1946; p. 275.

56. Verma, S. C.; McCalla, A. G. Cereal Chem., 1966, 43, 28.

57. Swanson, C. O.; Andrews, A. C. Cereal Chem., 1966, 22, 134.

58. Oka, S.; Babel, S. J.; Draudt, H. N. J. Food Sci., 1965,
    30, 212.

59. Barrett, F. F. In "Enzymes in Food Processing;" Reed, G.,
    Ed. Academic Press: N.Y., 1975; p. 301.

60. Beuchat, L. R. Cereal Chem., 1977, 54, 405.

61. Lynch, C. J.; Rha, C. K.; Catsimpoolas, N. J. Sci. Food
    Agric., 1977, 28, 971.

62. Blow, D. M.; Dirktoft, J. J.; Hartley, B. S. Nature,

    1969, 221, 337.

RECEIVED October 21, 1980.

Multiple Regression Modelling of Functionality

MAC R. HOLMES

Department of Agricultural Economics, University of Georgia,
Georgia Experiment Station, Experiment, GA 30212

Use of multiple regression techniques in the study of functional properties of food proteins is not new (1-6). Most food scientists have some familiarity with basic statistical concepts and some access to competent statistical advice. At least one good basic text on statistical modelling for biological scientists exists (7). A number of more advanced texts covering use of regression in modelling are available (8, 9).

The objectives of this paper are to present some potential uses of regression techniques in food protein research, to discuss some desirable steps in the modelling process, to present an example of the rationale underlying development of a model, and to discuss some potential statistical problems which might arise.

An Example of a Regression Model.

The basic structure for a regression model having two independent variables is as follows:

$$Y = a + b_1X_1 + b_2X_2 + e$$

where Y = the dependent variable,
a = the intercept, or constant, of the equation,
X_i = the ith independent variable,
b_i = the coefficient of X_i,
e = the variation in Y not explained by the preceding variables and coefficients; "e" is assumed to be normally distributed with mean of zero.

The model form implies that variations in X_1 and X_2 cause variations in Y but that some variation in Y is due to a random component (e) of Y. Since "e" has an expected value of zero, it is ordinarily not referred to in listing the estimated equation.

0097–6156/81/0147–0299$05.00/0

The objectives of estimating such an equation can be summarized as follows:

(1) To estimate the coefficients - a, b_1, b_2, for example - and thus the effects of variations in X_1 and X_2 (for example) on the level of Y, i.e. to estimate the response surface. In food protein research, the dependent variable could be a functional property of the protein material or some mathematical transformation thereof, and the independent variables (the X's) could be controllable conditions hypothesized to affect the functional property - such as pH, heat, or salt concentration - and/or mathematical transformations of these conditions. Throughout the remainder of this paper, the measured conditions which form the bases for the independent variables will be referred to as the factors.

(2) To estimate the statistical significance of the effects, i.e., the coefficients, of the independent variables on the dependent variable using t-tests of significance, to estimate the statistical significance of the overall model using the F-test, and to estimate the percentage of variation in the dependent variable explained by the equation using the coefficient of determination.

Development of a model must be based on a theory, or theories, concerning the effects of the factors on the functional property being studied. Such theories, or hypotheses, can be based on prior research results, theories developed by others, collection and preliminary analysis of data and, perhaps, intuition. In sum, the hypotheses are implied from what is already known or hypothesized. A prime requirement for use of regression is that there must be some way of objectively measuring levels of the functional property and of the factors in order to provide data to be used in estimation of the model coefficients.

The Experimental Design

To develop a model for a particular research experiment to study functional properties of a food protein, the researcher must obviously have some knowledge concerning which factors are potentially important determinants of the level of the functional property to be studied. The researcher must select 1) each of the factors for which there is to be some variation in level within the experiment, 2) the levels to be used for each variable factor, 3) the number of combinations of levels of the different factors to be used, and 4) the level at which each nonvariable, but potentially important, factor is to be set throughout the experiment.

These considerations will be primary determinants of the experimental design, which will determine the statistical precision and reliability of the models estimated.

The number of levels of any one factor which should be included in the experimental design will depend largely on the following considerations: 1) the research objectives (Is this factor of primary interest in this research?), 2) the degree of certainty attached to current knowledge of the effects of the factor, and 3) the degree of probability that this factor interacts with other factors, which are definitely of primary interest, to determine the levels of the functional property (or properties) to be studied. Setting the level of an important causative factor at an arbitrary point(s) could seriously bias results of analysis of the experimental data.

In setting the number of levels to be studied of any one variable factor, the type of effects which it is likely to have on the functional properties to be studied is very important. If its effects are known to be linear and that factor is of secondary importance to the researcher, then two levels (one at each end of some practical range of levels) may be sufficient. If, on the other hand, it is known that the effects of this factor are curvilinear and/or discontinuous at some point, then at least three levels should be included in the experimental design. If the interaction of a factor with other factors is known to be significant, then this too could be sufficient reason to include more than two levels of that factor in the design.

Multiple regression, also called ordinary least squares, can frequently provide reasonably precise, reliable estimates of coefficients even if the data analyzed are unbalanced, i.e., have "missing cells". However, all possible combinations of all levels of all variable factors included must be used if maximum statistical precision and reliability of the estimated coefficients is desired. If the experiment is very large (involving several levels of several factors), then it may be desirable to leave out some combinations. The researcher should weigh the costs of exclusion of combinations - loss of statistical precision and reliability - against the costs of inclusion of the combinations - usually time and money. This may best be done with the advice of a statistician.

One frequent problem in food science research is how to define a single observation for regression analysis (an observation is composed of a measurement of the dependent variable, before any transformation, studied along with the corresponding levels of the variable factors). An experimental unit is generally defined as the unit of material to which a single treatment (combination of levels of the variable factors) is applied; a sampling unit is that part of the experimental unit on which the effects of a treatment are measured (7, p. 90). Frequently, when chemical analyses are carried out to measure a functional property of a food protein material, multiple determinations will be made on each sampling unit and/or more than one sample will be drawn from each experimental unit. The problem is whether to include the results of each and every determination (or sample) as an observation or to average determinations (or samples) to obtain a single observation for each sampling (or experimental) unit. Three rules have been recommended for making such a choice (10). If the number of experimental units is small, if the coefficient of variation for each functional property measured within each experimental unit is small, and the determinations (or samples) are relatively homogeneous, then each determination (or sample) may be included in a separate observation; the values of the variable factors as set within each experimental unit are repeated for each observation based on that experimental unit. If, on the other hand, the number of experimental units is high and the coefficient of variation for each functional property measured within each experimental unit is high, then all of the values determined (or sampled) of the functional property within each experimental unit should be averaged to obtain one observation per experimental unit. Either method is recommended if the number of experimental units is large and the coefficients of variation within the experimental units are small. If the number of experimental units is small and the coefficients of variation within the experimental units are large, expand the number of experimental units and average the determination (or sample) values to obtain a single observation per experimental unit (10).

Examining the Data

Once an experimental design is selected and the data are collected, then preliminary analyses of the data should be carried out. Some simple statistics – such as means, ranges, standard deviations, etc. – should be calculated to familiarize the researcher with the data and to serve as bases for comparison with previous research. These statistics can also be useful when graphing equations estimated later.

A second type of analysis which can be extremely useful under some conditions is plotting of the dependent variable (functional property) data against the corresponding levels of each variable factor.

When only one variable factor is included in the experiment, then plotting of the resulting levels of the dependent

variable (or functional property) against the corresponding
variable factor levels will allow the researcher to see if
the data fall on a straight line, fall in a curvilinear
pattern, show any evidence of discontinuity around any level
of the variable factor, or indicate very little relationship
between levels of the functional property and those of the
variable factor (this would be indicated by an obvious scat-
tering of data points with no discernible patterns). Figure
1 shows a possible plot of points for an experiment. Figure
2 shows a curve which might fit these data in Figure 1 and
also indicates the form of the equation graphed in Figure 2,
$Y = a + b_1 X + b_2 X^2$. In this case, two independent variables, X and X^2, are formed from the one variable factor, X,
in order to fit a suitable nonlinear equation. However, it
should be noted that some nonlinear equation forms which may
be most suitable for some data are nonlinear in the coefficients; these cannot be estimated using ordinary regression techniques. These techniques have been discussed
elsewhere (8, 9).

One problem which may sometimes be most easily detected
using plots of the data is that of detecting "outliers", or
"bad" data points. These may have resulted from improper
application of experimental techniques, incorrect measurements, or other factors not accounted for in the experimental
design. Such data may be excluded from the regression analysis. However, care should be taken to not exclude legit-
imate data points arising from random variation in a functional property or from variation due to the consistent
influence of variable factors which should have been included
in the analysis (factors the influence of which could not
have been excluded).

However, when more than one variable factor is included
in an experiment, plotting of the functional property studied
against each of the variable factors may not be useful and
may indeed be misleading. Plotting of data points when there
are two or more variable factors is generally useful only
when there are several values of a variable factor for each
of one or more sets of fixed values of all other factors.
Otherwise, separation of the influences of different factors
on the functional property will usually be impossible on the
basis of plots of the data.

If plotting cannot be used in the selection of a proper
geometric representation (and the corresponding equation
form) of the influence of each variable factor being studied,
then accepted theory and past research results may aid the
researcher in selection of variable and equation forms to be
used. Within limits, the resuts of t-tests of regression
coefficients may be used to select the most appropriate
variables and variable transformations for use in the final

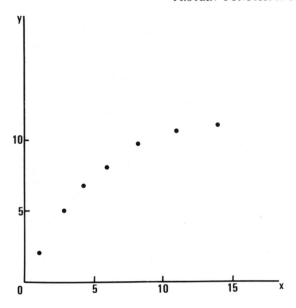

Figure 1.   Hypothetical plot of data points

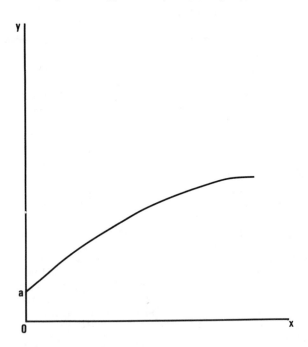

Figure 2.   Hypothetical graph of a quadratic equation

model. Thorough discussions of plotting, equation forms, variable transformations and selection of independent variables have been published elsewhere (8, Chapters 3-6; 9, Chapters 3-8).

An Application of Multiple Regression Techniques

At this point, it seems useful to examine an example of the application of multiple regression techniques to analysis of experimental data for which results have already been obtained and published.

McWatters and Holmes (4) developed multiple regression models of the effects of pH and salt concentration on functional properties of soy flour. Design of the experiment and selection of the factors to be included were based, in part, on earlier findings that emulsion capacity of defatted peanut meal was inhibited around the isoelectric point (ca pH 4.0) (2).

In the soy flour research, three levels of salt concentration (0.0, 0.1, and 1.0 M NaCl) were used, and nine levels of pH from 2.0 through 10.0 were used with all three levels of salt concentration. When measurements of emulsion capacity were plotted against pH within each level of salt concentration, a number of conclusions were evident. First, the data plots were radically different at the high salt concentration (1.0 M NaCl) from the plots at the lower salt concentrations (0.0 and 0.1 M NaCl). At the high salt concentration, the plot resembled a smoothly rising curve which flattened out at the high end of the pH range. But, at the lower salt concentrations, emulsion capacity tended to drop off sharply at pH levels around 4.0. Furthermore, the increases in capacity as pH was increased above 4.0 or decreased below 4.0 were curvilinear rather than linear as in Figure 3. Much the same effects were observed for emulsion viscosity and percent soluble nitrogen. As noted in the article, all of these effects were due to the combined ⌐ffects of salt concentration and pH on electrical charges on the particles in solution. Given these plots, regression equations were estimated which modelled the data very well. The equations are presented in Table 1 (Parts 1 and 2).

There are a number of key points to be made about the variables used in the equations. First, the equation forms used for the high salt concentration (1.0 M NaCl) are simple quadratic and cubic forms using pH and the square and cube of pH as independent variables. The high salt concentration negated the emulsion inhibiting effects of the isoelectric point. The percent of the variation in the function properties accounted for by these equations ranged from about 80 percent for emulsion viscosity to over 98 percent for emulsion capacity.

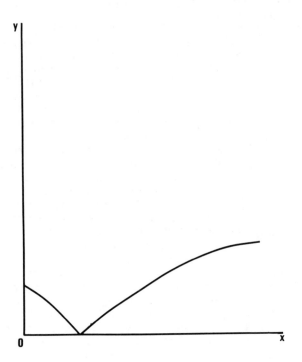

Figure 3.   Hypothetical graph of a discontinuous function

Table 1 (Part 1). Multiple regression models for prediction of nitrogen solubility and emulsifying properties of soy flour as influenced by pH and salt concentration.

Dependent variable with salt concentration	Variable description	Regr. coeffi.	Beta value	t-value[a]
% Soluble N_2 with 0.0 & 0.1 M NaCl (R^2 = 0.903; standard error of estimate = 10.107)	Constant	15.935		
	pH-4.0[b]	29.715	2.165	8.63**
	4.0-pH[c]	72.740	1.635	5.89**
	Salt level	-140.472	-0.237	-4.17**
	$(4.0-pH)^2$	-16.304	-0.691	-2.69**
	$(pH-4.0)^2$	-2.597	-1.075	-4.71**
% Soluble N_2 with 1.0 M NaCl (R^2 = 0.905; standard error of estimate = 4.659)	Constant	22.283		
	pH_2	15.658	6.830	6.83**
	pH^2	-0.917	-4.886	-4.89**
Emulsion capacity with 0.0 & 0.1 M NaCl (R^2 = 0.838; standard error of estimate = 0.081)	Constant	0.221		
	pH-4.0[b]	0.183	2.226	6.41**
	4.0-pH[c]	0.451	1.694	4.44**
	Salt level	-2.419	-0.682	-4.59**
	$(4.0-pH)^2$	-0.165	-1.166	-3.39**
	$(pH-4.0)^2$	-0.023	-1.585	-5.20**
	(Salt level)(pH-4.0)	0.592	0.642	4.00**
	(Salt level)(4.0-pH)	1.114	0.314	2.32*

a *Significant at 5% confidence level; ** significant at the 1% level.

b The variable described as "pH-4.0" is given a value of zero unless the pH is greater than 4.0.

c The variable described as "4.0-pH" is given a value of zero unless the pH is less than 4.0.

Source: McWatters, K. H., and Holmes, M. R. (4).

Journal of Food Science

Table 1 (Part 2). Multiple regression models for prediction of nitrogen solubility and emulsifying properties of soy flour as influenced by pH and salt concentration.

Dependent variable with salt concentration	Variable description	Regr. coeffi.	Beta value	t-value[a]
Emulsion capacity with 1.0 M NaCl (R^2 = 0.983; standard error of estimate = 0.010)	Constant	0.155		
	pH_2	0.108	21.020	21.02**
	pH^2	−0.007	−16.837	−16.84**
Emulsion viscosity with 0.0 and 1.0 M NaCl (R^2 = 0.785; standard error of estimate = 10,210.917)	Constant	4,701.905		
	$pH-4.0$[b]	18,031.429	1.908	5.18**
	$4.0-pH$[c]	117,147.143	3.825	9.40**
	$(4.0-pH)^2$	−47,409.048	−2.919	−7.74**
	$(pH-4.0)^2$	−2,127.619	−1.279	−3.82**
Emulsion viscosity with 1.0 M NaCl (R^2 = 0.796; standard error of estimate = 4,441.520)	Constant	98,207.619		
	pH_2	−28,811.400	−8.588	−3.47**
	pH_3	4,945.714	18.003	3.28**
	pH	−277.172	−10.463	−3.33**

a *Significant at 5% confidence level; ** significant at the 1% level.
b The variable described as "pH-4.0" is given a value of zero unless the pH is greater than 4.0.
c The variable described as "4.0-pH" is given a value of zero unless the pH is less than 4.0.

Source: McWatters, K. H., and Holmes, M. R. (4).

The models developed for the low salt concentrations (0.0 and 0.1 M NaCl) were very different from those for the high salt concentration. The two basic independent variables used were salt level and the absolute value of the pH minus 4.0. To avoid implying that the behavior of each functional property on either side of pH 4.0 is the mirror image of its behavior on the other side of pH 4.0, two variables were formed from the absolute value of pH minus 4.0. These were the absolute values of pH minus 4.0 for each pH above 4.0 (values of this variable for observations in which the pH was less than 4.0 were set equal to zero) and the absolute values of pH minus 4.0 for each pH below 4.0 (values of this variable for observations in which pH was greater than 4.0 were set equal to zero). As shown in the table, these basic variables were used in the estimated models along with their squares, their interactions with salt level (0.0 and 0.1 M NaCl), and salt level to form the independent variables in the final equations used. Other variables, such as cubic powers of pH minus 4.0, were tried and discarded due to lack of statistical significance in arriving at the final models.

A number of points should be noted concerning the statistics displayed in the table. First, if the researcher wishes to rank the variables in order of their importance within the equation, absolute values of the beta values are the appropriate indicators of rank (7, p. 284). Second, the t-values of the regression coefficients give us estimates of the statistical significance of the independent variables used. Third, the R-square, or coefficient of determination, is an estimate of the percent of variation in the dependent variable (the functional property) explained by the corresponding regression equation.

Some Potential Problems

A number of reservations concerning use of regression should be expressed. It is desirable that the experimental data be as balanced as possible (as close to having all combinations of all levels of all variable factors in the design as possible), though regression, also known as ordinary least squares, is a suitable technique for analyzing unbalanced data. Use of extremely unbalanced data may reduce the precision and/or reliability of the regression coefficients estimated particularly if the effects of the factors are not linear. A second consideration is that one of the assumptions of regression is that there is no significant correlation between the independent variables included in the model. Such correlation will exist if, as in the above equations, certain transformations of independent variables included in the model are also included as variables. This

will tend to increase the variance of the coefficients and
may, in some cases, affect estimation precision. One way of
reducing these problems is to code the data by subtracting
the mean of each basic factor from the value of the factor in
each observation and dividing the results by the standard
deviation of the factor (thus "standardizing" the variable;
11). Any transformations are then performed on the standardized data. Regression coefficients estimated using these
data must be decoded, however.

Uses of Estimated Models

A primary use of regression models is usually prediction
of the levels of the dependent variable (or functional property) under given conditions. Such predictions are most
reliable if the given conditions fall within the ranges of
the conditions included in the data used in estimation of the
model.

If the regression models are considered true behavioral
models of the functional properties (or good approximations
thereof) and reliable models (i.e., the R-square is generally
considered an indicator of reliability of a model), then yet
another use might be appropriate. If no interactions are
present, then the first derivative (if it exists for the
particular equation form) with respect to each factor (or
basic independent variable) may be taken as an estimate of
the marginal effect (effect of the last unit of the factor,
or basic variable, added) of that factor on the functional
property, or dependent variable. If a monetary value can be
placed on the dependent variable, then this estimated marginal effect can be multiplied times that monetary value to
obtain an estimate of the marginal revenue arising from a one
unit increase in the variable factor, or basic independent
variable. If a cost can be attached to the increase in the
level of the variable factor, then we can estimate the profit, or lack of profit, associated with an increase in the
level of the factor used by subtracting the marginal cost
from the marginal revenue.

If interactions exist, then partial derivatives must be
used rather than first derivatives. The major complication
arising here is that the partial derivative can be used as
specified above only by assuming that other variable factors
which appear in the partial derivatives are set at fixed
levels. Any competent production economist can assist in
setting up procedures designed to produce such results. A
number of references are available (12, 13, 14).

Conclusions

Multiple regression techniques can be used in modelling the effects of varying levels of environmental factors on functional properties of plant proteins. Use of multiple regression not only may allow the researcher to test statistical significance of properties, but it may also allow the researcher to estimate magnitudes of the effects of the environmental factors on behavior of functional properties. Multiple regression models have been found useful in estimating effects of such factors as moist heat, pH, and salt concentration on solubility and emulsifying properties of plant proteins. Some of these effects have been found to be non-linear and discontinuous. Use of the technique, however, is no substitute for good experimental design or knowledge of the data.

Literature Cited

1. Cherry, J. P.; McWatters, K. H.; Holmes, M. R. J. Food Sci., 1975, 40, 1199.
2. McWatters, K. H.; Cherry, J. P.; Holmes, M. R. J. Agric. Food Chem., 1976, 24, 517.
3. McWatters, K. H.; Holmes, M. R. J. Food Sci., 1979, 44, 765.
4. McWatters, K. H.; Holmes, M. R. J. Food Sci., 1979, 44, 770.
5. McWatters, K. H.; Holmes, M. R. J. Food Sci., 1979, 44, 774.
6. Sefa-Dedeh, S.; Stanley, D. J. Agric. Food Chem., 1979, 27, 1238.
7. Steel, R. G. D.; Torrie, J. H. Principles and Procedures of Statistics; McGraw-Hill, New York, 1960.
8. Daniel, C.; Wood, F. S. Fitting Equations to Data: Computer Analysis of Multifactor Data for Scientists and Engineers; Wiley, New York, 1971.
9. Draper, N. R.; Smith, H. Applied Regression Analysis; Wiley, New York, 1966.
10. Kubala, J. J.; Gacula, M.C.; Moran, M. J. J. Food Sci., 1974, 39, 209.
11. Snee, R. D. J. Quality Technology, 1973, 5 (2), 67.
12. Heady, E. O. Economics of Agricultural Production and Resource Use; Prentice-Hall, Englewood Cliffs, N.J., 1952.
13. Carlson, S. A Study on the Pure Theory of Production, Kelley and Willman, New York, 1956.
14. Allen, C. L. Elementary Mathematics of Price Theory; Wadsworth Pub. Co., Belmont, CA, 1962.

RECEIVED September 5, 1980.

INDEX

INDEX